Intelligent Cyber-Physical Systems Security for Industry 4.0

Intelligent Cyber-Physical Systems Security for Industry 4.0: Applications, Challenges and Management presents new cyber-physical security findings for Industry 4.0 using emerging technologies like artificial intelligence (with machine/deep learning), data mining, and applied mathematics. All these are essential components for processing data, recognizing patterns, modeling new techniques, and improving the advantages of data science.

Features

- Presents an integrated approach with Cyber-Physical Systems, CPS security, and Industry 4.0 in one place
- Exposes the necessity of security initiatives, standards, security policies, and procedures in the context of Industry 4.0
- Suggests solutions for enhancing the protection of 5G and the internet of things (IoT) security
- Promotes how optimization or intelligent techniques envisage the role of artificial intelligence-machine/deep learning (AI-ML/DL) in cyber-physical systems security for Industry 4.0

This book is primarily aimed at graduates, researchers, and professionals working in the field of security. Executives concerned with security management, knowledge dissemination, information, and policy development for data and network security in different educational, government, and non-government organizations will also find this book useful.

Chapman & Hall/CRC Computational Intelligence and Its Applications

Series Editor: Siddhartha Bhattacharyya

Intelligent Copyright Protection for Images
Subhrajit Sinha Roy, Abhishek Basu, and Avik Chattopadhyay

Emerging Trends in Disruptive Technology Management for Sustainable Development
Rik Das, Mahua Banerjee, and Sourav De

Computational Intelligence for Human Action Recognition
Sourav De, and Paramartha Dutta

Disruptive Trends in Computer Aided Diagnosis
Rik Das, Sudarshan Nandy, and Siddhartha Bhattacharyya

Intelligent Modelling, Prediction, and Diagnosis from Epidemiological Data: COVID-19 and Beyond
Siddhartha Bhattacharyya

Blockchain for IoT
Debarka Mukhopadhyay, Siddhartha Bhattacharyya, K Balachandran, and Sudipta Roy

Hybrid Intelligent Systems for Information Retrieval
Anuradha D Thakare, Shilpa Laddha, and Ambika Pawar

Intelligent Cyber-Physical Systems Security for Industry 4.0: Applications, Challenges, and Management
Jyoti Sekhar Banerjee, Siddhartha Bhattacharyya, Ahmed J Obaid, and Wei-Chang Yeh

For more information about this series, please visit:
https://www.crcpress.com/Chapman--HallCRC-Computational-Intelligence-and-Its-Applications/book-series/CIAFOCUS

Intelligent Cyber-Physical Systems Security for Industry 4.0

Applications, Challenges and Management

Edited by
Jyoti Sekhar Banerjee
Siddhartha Bhattacharyya
Ahmed J Obaid
Wei-Chang Yeh

CRC Press
Taylor & Francis Group
Boca Raton London New York

CRC Press is an imprint of the
Taylor & Francis Group, an **informa** business
A CHAPMAN & HALL BOOK

First edition published 2022
by CRC Press
6000 Broken Sound Parkway NW, Suite 300, Boca Raton, FL 33487-2742

and by CRC Press
4 Park Square, Milton Park, Abingdon, Oxon, OX14 4RN

CRC Press is an imprint of Taylor & Francis Group, LLC

Library of Congress Cataloging-in-Publication Data
Names: Banerjee, Jyoti Sekhar, editor.
Title: Intelligent cyber-physical systems security for industry 4.0:
applications, challenges and management / Jyoti Sekhar Banerjee,
Siddhartha Bhattacharyya, Ahmed J Obaid, Wei-Chang Yeh.
Description: Boca Raton: Chapman & Hall/CRC Press, 2022. | Series: Chapman &
Hall/CRC computational intelligence and its applications series
| Includes bibliographical references and index.
| Identifiers: LCCN 2022031345 (print) | LCCN 2022031346 (ebook)
| ISBN 9781032148342 (hardback) | ISBN 9781032148359 (paperback)
| ISBN 9781003241348 (ebook)
Subjects: LCSH: Cooperating objects (Computer systems) | Industry 4.0.
Classification: LCC T59.6 .I58 2022 (print) | LCC T59.6 (ebook)
| DDC 658.4/038028563--dc23/eng/20220919
LC record available at https://lccn.loc.gov/2022031345
LC ebook record available at https://lccn.loc.gov/2022031346

ISBN: 978-1-032-14834-2 (hbk)
ISBN: 978-1-032-14835-9 (pbk)
ISBN: 978-1-003-24134-8 (ebk)

DOI: 10.1201/9781003241348

Typeset in Palatino
by SPi Technologies India Pvt Ltd (Straive)

Jyoti Sekhar Banerjee would like to dedicate this book to his dear parents, wife (Arpita), and loving Son (Rishik).

Siddhartha Bhattacharyya would like to dedicate this book to his loving wife Rashni.

Ahmed J Obaid would like to dedicate this book to his family and many friends. A special feeling of gratitude to his loving wife Zainab R. Raheem, son Hamzah, daughters Fatema, Malak, Teba, and Basmalah. He also dedicates this work to the University of Kufa, Faculty of Computer Science and Mathematics.

Wei-Chang Yeh would like to dedicate this book to the team who brought this book to life and would like to thank the team for their unparalleled brilliance.

Contents

Preface

In recent years, we have witnessed an exponential growth in the development and deployment of various types of cyber-physical systems (CPS). They have brought impacts to almost all aspects of our daily life, for instance, in electrical power grids, oil and natural gas distribution, transportation systems, healthcare devices, household appliances, and many more. Many of such systems are deployed in the critical infrastructure and life support devices or are essential to our daily lives. Therefore, they are expected to be free of vulnerabilities and immune to all types of attacks, which, unfortunately, is practically impossible for all real-world systems.

With the exponential growth of CPS, new security challenges have emerged. Various vulnerabilities, threats, attacks, and controls have been introduced for the new generation of CPS. However, there is a lack of a systematic study of CPS security issues. In particular, the heterogeneity of CPS components and the diversity of CPS systems have made it very difficult to study the problem with a narrow span.

Both intelligent cyber-physical systems and Industry 4.0 are the most promising fields of research nowadays. Security is perhaps the biggest challenge to maintain systems secure in this digital world. The fourth industrial revolution (Industry 4.0) is closely related to intelligent CPS security issues. This book introduces new outcomes in intelligent CPS security for Industry 4.0, its possible applications, challenges, and management for different domains, and their socio-economical influences on the community. Industry 4.0 demands different propositions in the context of secure connections, control, and maintenance of the intelligent CPS and intensifying their intercommunication with people through connected devices.

This volume presents new, reliable solutions for ensuring the cyber-security of industrial CPS in Industry 4.0. This volume also focuses on the types of cyber-physical attacks, attack models, security policies and procedures, architectural approach to system design, and integration in the context of industry 4.0. The volume comprises 12 contributory chapters apart from the introductory and concluding chapters to report the latest developments in this direction.

National essential infrastructures such as home automation systems, transportation networks, electricity grids, and IoT-enabled other services rely on CPS. Again, CPS is an integral part of Industry 4.0. Due to its widespread usage, CPS should be given high priority for security purposes in these applications. If compromised, these systems, which are part of crucial infrastructure, would have disastrous results. In order to protect the CPS, an in-depth knowledge of the security dangers, vulnerabilities, and assaults is needed. Security threats, solutions, and challenges for CPS have been examined in Chapter 1. Current security markets and security problems provide a variety of hurdles to the implementation of CPS. A look into CPS vulnerabilities, security analysis, and attacks is also included in the research. As the last step, the authors offer remedies for each system of CPS security risks, as well as outline possible answers to future difficulties.

The advent of augmented reality (AR) systems has radically altered computing devices and current technological trends. Following this revolution, there is still a big issue of proper authentication with these systems. Authentication is the primary defense when it comes to secured and authorized access to a system. Currently, authentication in AR systems is being done via smartphones or personal laptops and is sometimes neglected also. In Chapter 2, the authors describe the issues with several authentication protocols which are being currently used by the AR systems and other proposed mechanisms. We have compared different protocols in detail and categorized the issues based on their security against attacks and cryptography/algorithm used. This book also highlights the integration of AR with CPS for prototyping purposes and aims to provide a better understanding of the existing protocols to develop or update existing protocols in the future which are secured and difficult to compromise.

With the abundance of digital text in social media, text steganography has become a lucrative approach for hidden communication. Different text steganography methods have been proposed in recent days but nothing is found for automation of existing manual cover text selection. Chapter 3 presents an automated cover selection system with an aim of faster stego-text generation. The different attributes of secret texts along with the requirement study of chosen stego-algorithms with the application of fuzzy clustering helps in finding the most accurate cover text from a pre-collected cover-text-bank. An interactive system has been designed in this regard, considering stego-algorithms and secret texts as input. A threshold secret text vector is generated by analyzing the inputs and cover text vectors are also obtained from each cover text of the cover-text-bank. The fuzzy clustering method finally finds the closest cover text that is able to hide the chosen secret text using the chosen stego-algorithm. The proposed method is experimentally tested and verified to be an accurate and effective one.

In the emerging internet of things (IoT) era, billions of smart devices and machines are seamlessly interconnected with each other over the internet to exchange information and support decision-making. The IoT is progressively becoming an important aspect of the fourth industrial revolution (often referred to as industrial internet of things [IIoT] or Industry 4.0) and provides an unprecedented opportunity to revolutionize traditional production and manufacturing processes. To achieve the goals and realize the vision of IIoT, a communication protocol specifically designed for low-power and lossy networks, a.k.a. RPL, has stood out from the crowd and quickly became a promising routing protocol for the IIoT. However, the security and privacy issues were not the major concerns when RPL was designed. Thus, it fails to meet the security requirements of IIoT. In Chapter 4, the authors first introduce RPL routing protocol and its major component, Trickle algorithm. Then, they identify and analyze various RPL-specific attacks in the IIoT, discuss their corresponding countermeasures, and present their performance impact via preliminary simulation results. Finally, the chapter concludes with future research directions, including interdisciplinary aspects and insights.

An online cyber-attack can have similar real-world consequences as in the digital world and this is not a new concept. The number of cyber-physical attacks that target the physical world is increasing rapidly nowadays. Since the arrival of the concept of internet of everything, the number of smart devices connecting to the internet has increased exponentially. With this increase, the "attack surface" has increased as well. A cyber-physical attack is any cyber-attack or security breach that has a direct impact on the physical world. A hacker can hack into the system and gain control of the various IoT devices installed by the users or businesses and carry out malicious activities and even cause damage to properties and even put lives at risk. In Chapter 5, the authors highlight a number of cyber-physical attacks that are happening in the IoT environment and some of the measures to, if not eliminate, reduce them significantly and make them sustainable.

In the data communication system, security must be concentrated. Security and the complexity of algorithms dealing with security increase because of the randomization of secret keys. Hence, recently, cryptography algorithms must be capable of balancing enormous memory and execution time based on the hardware platform. FPGAs (field-programmable gate arrays), a reprogrammable device, are attractively used to implement hardware employed in encryption algorithms. Advanced Encryption Standard (AES), an efficient and cost-effective symmetric cryptographic algorithm, is broadly used in applications like ATMs, smart cards, mobiles, and internet servers to maintain data confidentiality. A secure, robust cryptosystem depends on AES Substitution Box, and chaotic components are proposed in Chapter 6, which comprise an effective Pseudo-Chaotic Number Generator (PCNG), global diffusion and block cipher. The finite field is used to define the PCNG, which eliminates the danger of depreciated security as a subsequent dynamic degradation. At the same time, numerical implementation of chaotic maps is always stated as real numbers. In the modified Bernoulli process, diffusion properties are increased efficiently by horizontal addition diffusion (HAD) and vertical addition diffusion (VAD). The strength of cryptography is tested in detail in S-Box proposed through the various benchmark standards. The results of experiments and examination of performance showed that this chaos-based advanced encryption algorithm proposed using S-box Bernoulli process (CA-Sbox-BP) achieves 96% efficiency, 64% non-linear criterion, and 41% strict avalanche effect with 98 Kbps in 75 ms.

The overall increase in CPS has led to the emergence of new security challenges. Numerous vulnerabilities, threats, attacks, and controls have been implemented for the future version of CPS. The rapid and considerable rise of CPS has had an impact on the general well-being of people and encouraged the creation of a relatively large variety of services and applications, including electronic healthcare, smart homes, and electronic commerce. Interconnecting the virtual and real worlds, on the other hand, presents new and critical security challenges. However, there is a scarcity of systematic reviews of CPS security research. The heterogeneity of CPS components, as well as the diversity of CPS systems, has hindered attempts to investigate the problem using a single generalized model. Chapter 7 looks at the various forms of attacks that can be performed in CPSs, as well as mitigation approaches and attack defensive strategies. Existing security approaches are also provided and examined, and their key limitations are discussed. Finally, a few suggestions and recommendations are made in the context of the key findings throughout this in-depth review.

Human beings are living in a digital world. Technologies like big data, cloud computing, IoT, artificial intelligence (AI), smart consumer electronics, etc. make life comfortable and save a lot of the human labor force. But, the embedded vulnerabilities, hazards, and loopholes of these technologies sometimes make life alarming with the recent increase in cyber-attacks and uncontrollable threats. This is quite evident in the lacking of traditional cyber security technologies, and these challenges bring up new technology that can solve the global need in the purview of cyber technology and social cyber awareness. Cyber security is one of the prominent fields of research as human beings are day by day shifting all kinds of operations into the cyber world. Industry 4.0 is fully dependent upon the autonomous process of digitalization. Hence, security is the first and foremost priority in designing any kind of system. Chapter 8 would research CPS, current cyber security issues, cyber-physical attacks, and their solving aspect emphasizing mostly deep learning and machine learning approaches. This introspection helps fill up the gaps that come across while dealing with any cyber security issues more confidently and in a speedy manner.

A CPS should be designed in such a way that it can face more than one type of attack. In Chapter 9, the authors introduce the characteristics of a CPS and how the attacker can probe the weakness of the cyber-physical ecosystem. The authors carry out a comprehensive literature survey, representing all the research work carried out to develop threat models and their areas of applicability. They propose the application of temporal logic to develop threat models which include the Prior's theory of branching-time model, the Peircean branching-time temporal logic model, and the Ockhamist branching-time temporal logic model. The experimental results are presented regarding how the threat can be mitigated even if the attacker tampers with the actuators or sensors in a CPS. The experimental results also show the satisfactory performance of the proposed models. In the future, hybrid models can be applied to develop threat models to challenge the CPS even more and how to counter such an attacker's threat. The chapter is concluded by highlighting the benefits of the proposed threat models and the amount of leverage they will provide to the designers of CPS security.

SDN, or software-defined networking, is a novel networking paradigm that has grabbed the interest of academics, researchers, and businesses alike. The fundamental network architecture is separated from applications, and network intelligence and state are theoretically centralized. SDN enhances network security by giving global knowledge of network status and enables disputes to be resolved rapidly from the logically centralized control plane. Although SDN is undergoing rapid development, it is still far from being deemed safe and trustworthy, which hinders its quick adoption in diverse network situations. In order to bridge the gap between SDN acceptance and deployment, the scientific community has been pushed to research the domain of SDN security in recent years. As more SDN devices and systems become available, security in SDN has to become a primary consideration. Chapter 10 covers a set of fundamental ideas and principles in order to provide a quick overview of the SDN architecture and the OpenFlow protocol and various challenges in SDN networks. It also emphasizes key SDN security challenges and includes a list of the most prevalent attacks detected in each layer of the SDN architecture, as well as an overview of various security solutions. This also presents a brief overview of SDN in CPS.

The term Industry 4.0 incorporates a guarantee of another modern upheaval: one that progresses fabricating procedures with the internet of things to make the world more digital and advanced. The term Industry 4.0 along with the internet of things helps to fabricate interconnected frameworks, which convey, break down, and use data to drive further shrewd activity back in the physical world, and also impart, examine, and use the data to drive further savvy activity back in the actual world. CPS and machine learning (ML) applications are generally utilized in Industry 4.0, advanced mechanics, and actual security. Improvement of ML strategies in CPS is unequivocally connected to the meaning of CPS. For the most part, ML assists CPS with learning and adjusting by utilizing smart models that are created from preparing enormous scope of information in the wake of handling and examination. Chapter 11 concludes the study of AI and ML based on security-related fields and sectors and also gives the idea to enhance collaboration among diverse departments, giving the right statistics to be handed to the proper human beings on a real-time foundation.

Avionics systems are the gauging instruments and flight control and monitoring devices employed on the aircraft, space vehicles, rockets, and satellites for successfully navigating the flight throughout the journey time. For the successful retention of a safe flight environment, the avionics must be assiduously steadfast and thoroughly functional. Aerospace industries are rapidly moving toward the incorporation of AI-enabled systems into avionics for the sake of upgrading the flight control processes to attain higher reliability and error-free computing. These CPS require new AI algorithms based on punctilious mathematical modeling and allied aerospace engineering simulations. Thus, they help achieve highly scalable prototyped solutions capable of handling real-time problems encountered during critical flight phases. Chapter 12 introduces the triangulation augmented AI algorithm composed of a series of novel mathematical models that could be effectively put into practice for the efficient enablement of artificially intelligent systems that ultimately supplement the management of spacecraft avionics systems for executing attitude stabilization criteria on a vehicle suffering from major disturbances leading to the non-effectuation of its auto-stabilizing conditions. Following the different functional modules, triangulation-based mathematical modeling for achieving enhanced cyber-physical avionics functionalities, that account for improved vehicle performance, is incepted that makes use of a triangulation determining mathematical formula.

Two or more automobiles can communicate with each other in a vehicular communication network (VCN). The security of messages sent between cars, on the other hand, is critical. We must guard against the misuse of data from various types of beacons that arrive during the communication process. In Chapter 13, the authors discuss various forecasting-based authentication schemes in VCN, including privacy-based authentication, elliptic curve digital signatures, light-based authentication schemes, blockchain-based security authentication scheme, conditional privacy-preserving authentication scheme, position-based prediction technique, and many others. This approach assists in anticipating various security attacks. This scheme can be used to analyze packet loss and to set up beacons in advance for proper vehicle communication. These methods provide more secure and effective communication between vehicles, as well as better network resource management.

Finally, Chapter 14 presents the concluding remarks on the attributes of a secure CPS with reference to the need for computational intelligent algorithms in this regard.

This volume explains the need to implement cloud computing and big data more broadly while addressing the Industry 4.0 requirements for huge data flows generated by swarms of IIoT devices. It is also necessary to introduce technologies and concepts for Industry 4.0 vertical and industrial integration, such as semantic service oriented architecture (SSOA), enabling a digital twin; however, these are required for seamless data integration and contextualization for data modeling and device representation based on unique identifiers. The primary audience of this volume includes research students of computer science and information technology apart from the software professionals who can inherit the much-required developmental ideas to boost their knowledge in these related fields.

Kolkata, India   **Jyoti Sekhar Banerjee**
Birbhum, India   **Siddhartha Bhattacharyya**
Kufa, Iraq   **Ahmed J Obaid**
Hsinchu, Taiwan   **Wei-Chang Yeh**
May 2022

Editors

Jyoti Sekhar Banerjee, B.Tech, M.E, Ph.D. (Engg.), is currently serving as the Head of the Department in the Computer Science and Engineering (AI & ML) Department at the Bengal Institute of Technology, Kolkata, India and visiting researcher (Post Doc) at Nottingham Trent University, UK. In addition, he is also the Professor-in-Charge, R & D and Consultancy Cell of BIT. He has teaching and research experience spanning 18 years and completed one IEI-funded project. He is a life member of the CSI, IEEE, ISTE, IEI, ISOC, and IAENG and fellow of IETE. He is the present honorary Secretary-cum-Treasurer of the ISTE WB Section. He is also the present honorary Secretary of the Computer Society of India, Kolkata Chapter. He is also the Execom Member of the IETE, Kolkata Centre. He has published over 50 papers in various international journals, conference proceedings, and book chapters. He is the lead author of "A Text Book on Mastering Digital Electronics: Principle, Devices, and Applications". He also filed two Indian patents. He has also co-authored another book and is currently processing six edited books with reputed international publishers like Springer, CRC Press, De Gruyter, etc. Presently, he is also processing two more textbooks; those are now in press. Currently, he is serving as the General Chair of GCAIA 2021, 2023, ICHCSC 2022, 2023. Dr. Banerjee served as a Guest Editor of ICAUC_ES 2021 and ICPAS-2021 issues in IOP journal of Earth and Environmental Science, Physics, Scopus indexed proceeding and in JESTEC journal (Scopus, WOS). His areas of research interests include computational intelligence, cognitive radio, sensor networks, AI/ML, network security, different computing techniques, IoT, WBAN (e-healthcare), and expert systems.

Siddhartha Bhattacharyya did his Bachelors in Physics, Bachelors in Optics and Optoelectronics, and Masters in Optics and Optoelectronics from the University of Calcutta, India, in 1995, 1998, and 2000, respectively. He completed his Ph.D. in Computer Science and Engineering from Jadavpur University, India, in 2008. He is the recipient of the University Gold Medal from the University of Calcutta for his Masters. He is the recipient of several coveted awards, including the Distinguished HoD Award and Distinguished Professor Award conferred by the Computer Society of India, Mumbai Chapter, India, in 2017, the Honorary Doctorate Award (D. Litt.) from the University of South America, and the South East Asian Regional Computing Confederation (SEARCC) International Digital Award ICT Educator of the Year in 2017. He has been appointed as the ACM Distinguished Speaker for the tenure of 2018–2020. He has been inducted into the People of ACM hall of fame by ACM, the USA, in 2020. He has been appointed as the IEEE Computer Society Distinguished Visitor for the tenure of 2021–2023. He has been elected as a full foreign member of the Russian Academy of Natural Sciences and the Russian Academy of Engineering. He is the founding President of the Kolkata Branch of the Asia-Pacific Artificial Intelligence Association (AAIA), Hong Kong.

He is currently serving as the Principal of Rajnagar Mahavidyalaya, Rajnagar, Birbhum. He served as a Professor in the Department of Computer Science and Engineering of Christ University, Bangalore. He served as the Principal of RCC Institute of Information Technology, Kolkata, India, during 2017–2019. He has also served as a Senior Research Scientist in the Faculty of Electrical Engineering and Computer Science of VSB Technical

University of Ostrava, Czech Republic (2018–2019). Prior to this, he was the Professor of Information Technology at RCC Institute of Information Technology, Kolkata, India. He served as the Head of the Department from March 2014 to December 2016. Prior to this, he was an Associate Professor of Information Technology at RCC Institute of Information Technology, Kolkata, India, from 2011 to 2014. Before that, he served as an Assistant Professor in Computer Science and Information Technology at the University Institute of Technology, The University of Burdwan, India, from 2005 to 2011. He was a Lecturer in Information Technology at Kalyani Government Engineering College, India, during 2001–2005. He is a co-author of six books and the co-editor of 88 books and has more than 400 research publications in international journals and conference proceedings to his credit. He has got two patents under the Patent Cooperation Treaty (PCT) and 19 patents to his credit. He has been a member of the organizing and technical program committees of several national and international conferences. He is the founding Chair of ICCICN 2014, ICRCICN (2015, 2016, 2017, 2018), and ISSIP (2017, 2018) (Kolkata, India). He was the General Chair of several international conferences like WCNSSP 2016 (Chiang Mai, Thailand), ICACCP (2017, 2019; Sikkim, India), ICICC 2018 (New Delhi, India), and ICICC 2019 (Ostrava, Czech Republic).

He is the Associate Editor of several reputed journals, including Applied Soft Computing, IEEE Access, Evolutionary Intelligence, and IET Quantum Communications. He is the editor of the *International Journal of Pattern Recognition Research* and the founding Editor in Chief of the *International Journal of Hybrid Intelligence*, Inderscience. He has guest-edited several issues with several international journals. He is serving as the Series Editor of IGI Global Book Series Advances in Information Quality and Management (AIQM), De Gruyter Book Series Frontiers in Computational Intelligence (FCI), CRC Press Book Series(s) Computational Intelligence and Applications & Quantum Machine Intelligence, Wiley Book Series Intelligent Signal and Data Processing, Elsevier Book Series Hybrid Computational Intelligence for Pattern Analysis and Understanding, and Springer Tracts on Human Centered Computing.

His research interests include hybrid intelligence, pattern recognition, multimedia data processing, social networks, and quantum computing.

He is a life fellow of the Optical Society of India (OSI), India, a life fellow of the International Society of Research and Development (ISRD), a full fellow of the The Royal Society for Arts, Manufacturers and Commerce (RSA), London, UK, a fellow of the Institution of Engineering and Technology (IET), UK, a fellow of Institute of Electronics and Telecommunication Engineers (IETE), India, and a fellow of Institution of Engineers (IEI), India. He is also a senior member of the Institute of Electrical and Electronics Engineers (IEEE), USA, International Institute of Engineering and Technology (IETI), Hong Kong, the Association for Computing Machinery (ACM), USA, and the Asia-Pacific Artificial Intelligence Association (AAIA), Hong Kong.

He is a life member of the Cryptology Research Society of India (CRSI), Computer Society of India (CSI), Indian Society for Technical Education (ISTE), Indian Unit for Pattern Recognition and Artificial Intelligence (IUPRAI), Center for Education Growth and Research (CEGR), Integrated Chambers of Commerce and Industry (ICCI), and Association of Leaders and Industries (ALI). He is a member of the International Rough Set Society (IRSS), International Association for Engineers (IAENG), Hong Kong, Computer Science Teachers Association (CSTA), USA, International Association of Academicians, Scholars, Scientists and Engineers (IAASSE), USA, Institute of Doctors Engineers and Scientists (IDES), India, The International Society of Service Innovation Professionals (ISSIP), and The Society of Digital Information and Wireless Communications (SDIWC). He is also a certified Chartered Engineer of the Institution of Engineers (IEI), India. He is on the Board of Directors of the International Institute of Engineering and Technology (IETI), Hong Kong.

Ahmed J Obaid is a full assistant professor at the Department of Computer Science, Faculty of Computer Science and Mathematics, University of Kufa. He received his BSC in Information Systems (IS) in 2005, from faculty of computer science, University of Anbar in 2005, M. Tech in Computer Science and Engineering from SIT, JNTUH, India, in 2012, and PhD in Web Mining and Data Mining from University of Babylon in 2017. His main line of research is web mining techniques and application, image processing in the web platforms, image processing, genetic algorithm, and information theory. Ahmed J. Obaid is an Associate Editor in *Brazilian Journal of Operations & Production Management* (ESCI), Guest Editor in KEM (Key Engineering Material, Scopus) Journal, Guest Editor of MAICT-19 and ICMAICT-20 issues in IOP journal of Physics, Scopus indexed proceeding, Guest Editor in JESTEC journal (Scopus, WOS) Journal, Managing Editor in *American Journal of Business and Operations Research* (AJBOR), USA, and Associate Editor in IJAST Scopus Journal. Ahmed J. Obaid is also a Reviewer in many Scopus Journals (Scientific publication, Taylor and Francis, ESTA, and many others). He is leader of ICOITES, MAICT-19, MAICT-20 EVENTS. Ahmed J. Obaid has supervised several final projects of Bachelor and Master in his main line of work. He has edited some books, such as Advance Material Science and Engineering in Scientific.net publisher, he has authored and co-authored several scientific publications in journals and conferences, and is a frequent reviewer of international journals and international conferences.

Wei-Chang Yeh is currently a distinguished professor of the Department of Industrial Engineering and Engineering Management at National Tsing Hua University in Taiwan. He received his M.S. and Ph.D. from the Department of Industrial Engineering at the University of Texas at Arlington. The majority of his research is focused around algorithms, including exact solution methods and soft computing. He has published more than 350 research papers in highly ranked journals and conference papers and has been awarded the Outstanding Research Award twice, the Distinguished Scholars Research Project once, and an Overseas Research Fellowship twice by the Ministry of Science and Technology (MOST) in Taiwan. He has been invited to serve as an Associate Editor of the three journals, namely the *IEEE Transactions on Reliability*, *IEEE Access*, and *Reliability Engineering & System Safety*. He proposed two novel algorithms called the simplified swarm optimization (SSO) and binary-addition-tree algorithm (BAT) that demonstrated the simplicity, effectiveness, and efficiency of his SSO and BAT for solving NP-hard problems. He has been granted 54 patents, enlisted on the global list of the top 2% of scientists (2020) by Stanford University, and earned an International Fellow, including the MOST fellow, the Guoguang Invention Medal, and the titles of Outstanding Inventor of Taiwan and Doctor of Erudition by the Chinese Innovation and Invention Society.

Contributors

Ankita Agarwal
Babu Banarasi Das Institute of Technology
    and Management
Lucknow, India

Vartika Agarwal
Graphic Era (Deemed to be University)
Dehradun, Uttarakhand, India

Mehtab Alam
Jamia Hamdard
New Delhi, India

Chirag R. Anand
Harvard University
Cambridge, Massachusetts, United States

Aneke chikezie Samuel
University of Uyo
Akwa Ibom State, Nigeria

Philip Asuquo
University of Uyo
Akwa Ibom State, Nigeria

Afolabi Awodeyi
University of Uyo
Akwa Ibom State, Nigeria

Ajay Kumar Balmiki
Maulana Abul Kalam Azad University of
    Technology
Kolkata, West Bengal, India

Jyoti Sekhar Banerjee
Bengal Institute of Technology
Kolkata, India

Siddhartha Bhattacharyya
Rajnagar Mahavidyalaya
Birbhum, India

Bibhorr
IUBH International University
Mulheimer Str. 38, Bad Honnef, Germany

Somenath Chakraborty
The University of Southern Mississippi
Hattiesburg, MS, USA

A. S. N. Chakravarthy
JNTUK
Vizianagaram, India

Suvamoy Changder
National Institute of Technology
Durgapur, India

Joyati Chattopadhyay
Techno International New Town
Kolkata, India

Kuntala Das
Bengal Institute of Technology
Kolkata, India

Sima Das
Maulana Abul Kalam Azad University
    of Technology
Kolkata, India

Ihtiram Raza Khan
Jamia Hamdard
New Delhi, India

Anandaprova Majumder
Dr. B. C. Roy Engineering
    College
Durgapur, India

Kaushik Mazumdar
IIT Dhanbad (ISM)
India

Roohie Naaz Mir
National Institute of Technology
Srinagar, India

V. Nandan
Vel Tech Rangarajan Dr Sagunthala R & D
    Institute of Science and Technology
Tamil Nadu, India

Hemlata Pant
Maharishi University of Information
    Technology
Lucknow, India

Cong Pu
Marshall University
United States

R. Gowri Shankar Rao
Vel Tech Rangarajan Dr Sagunthala R&D
    Institute of Science and Technology
Tamil Nadu, India

Pranjali Shah
Babu Banarasi Das Institute of Technology
    and Management
Lucknow, India

Sachin Sharma
Graphic Era (Deemed to be
    University)
Dehradun, India

Aman Srivastava
Babu Banarasi Das Institute of Technology
    and Management
Lucknow, India

Bliss Stephen
University of Uyo
Akwa Ibom State, Nigeria

Midighe Usoh
University of Uyo
Akwa Ibom State, Nigeria

Manas Kumar Yogi
JNTUK
Kakinada, India

Omerah Yousuf
National Institute of Technology
Srinagar, India

1

Intelligent Cyber-Physical Systems Security for Industry 4.0: An Introduction

Kuntala Das and Jyoti Sekhar Banerjee
Bengal Institute of Technology, Kolkata, India

Joyati Chattopadhyay
Techno International New Town, Kolkata, India

Siddhartha Bhattacharyya
Rajnagar Mahavidyalaya, Birbhum, India

CONTENTS

1.1 Introduction

Our physical and online worlds have come together in a new system known as the cyber-physical system (CPS). In terms of balance and level, it is strongly integrated with a variety of cyber and physical systems. An atmosphere of digital computation, discussion, and supervision is known as the "cyber environment" in the CPS. When it comes to internet of things (IoT) and sensors, the physical environment is a constant stumbling block. Human–machine interfaces and other ideas are all tied to the CPS's embedded systems, actuators, sensors, hardware, software, and other components. In order to collect, process, calculate, and analyze information about the physical atmosphere, a network of control devices, actuators, and sensors is used. The network's outputs are then applied to the physical environment. In the context of the IoT, CPS is a system-based dispersed control system that integrates a real-world device with actuators and sensors as well as a computational component that regulates it. There is a great deal of interconnectedness amidst the physical and cyber worlds [1] because of the rapid rise in information and communication technologies [2]. Manufacturing, medical, transportation, and energy sectors are all relying more and more on the CPS.

We might think of CPS as a form of network that connects physical and cyber systems. In large-scale initiatives, CPS integrates interconnected processing resources into the physical process to provide new capacity to the unique system and experience contemporaneous insight, data services, and active control. In today's world, CPS is used in various regions like electric power, chemical engineering, steel industries, nuclear amenities, etc., that are intertwined with the people's jobs and national economy [2] (Figure 1.1).

The application, data transmission, and perceptual layers are the three tiers of the CPS [3]. In the first layer, known as the perception layer, we have the IoT, camera, actuator, sensor, radio-frequency identification (RFID), and global positioning system (GPS).

As a result of side-to-side node collaboration, sensor data may be generated immediately in both local network domains and wide areas. Accordingly, data are gathered and sent to the communication layer via the perception layer, which then connects the IoT devices in the network [2, 4]. The sensor and application exchange data in communication, which is handled by the communication layer. Router, switch, wide area network (WAN), and local area network (LAN), 5G, 4G, WiFi, ZigBee, and Bluetooth are all used by this layer to communicate. CPS traditionally has a broad spectrum for global and local considerations, which is one of its major components. Real-time (RT) transmission reliability and hold-ups are both the responsibility of the communication layer. Depending on the request, the application layer might be referred to by a number of different names and be applied to a variety of fields.

This research explores CPS from the aspect of security at several levels, using a cross-layer methodology. CPS security may be separated into three tiers, again from a cross-layer viewpoint [5].

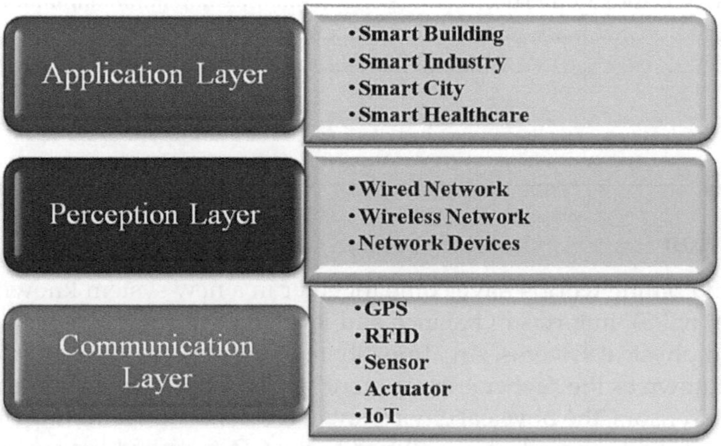

FIGURE 1.1
Three layers of CPS.

1. Hardware security mechanisms in CPS are another important part of the system's safety. Backdoors and trojans may damage essential hardware components, making security protection systems useless at the circuit or system level. As a second option, protection-enabled hardware may assist protect CPS by delivering efficient and effective mechanisms.

2. While intelligent devices make up the core of a CPS system, they are often overlooked when it comes to safety. As a consequence of this approach, devices are being produced without enough safety considerations in mind. There is a lot of anxiety about the security of industrial and commercial smart appliances nowadays. In addition, design solutions are advised to increase the robustness of smart devices.

3. Home automation will be a major component of future smart grid deployments because of its importance in our daily lives. Home automation systems are vulnerable to a variety of dangers, and this section provides an overview of threat strategies and countermeasures.

1.1.1 Smart Home Security

Particularly in the suburban side, the smart house has emerged as a critical component of the smart grid. Advanced metering infrastructure (AMI) requires smart home structures to include controls and schedules to facilitate the monitoring of household activities in an effort to conserve energy. Because there are so many people living on the grid, even modest reductions in energy usage may have a significant impact [5].

1.1.2 Cyberattacks of the Smart Home

The topic of smart home cybersecurity has piqued a great deal of curiosity among researchers. Hackers may use hardware backdoors to launch cyberattacks. This is not the first time that smart gadget flaws have been publicized. One recent example of this is the Google Nest thermostat, which was shown to be vulnerable. Intruders could be able to control the Nest thermostat from afar if it is set up that way. Cyberattacks against smart gadgets are, in fact, a frequent topic in the news. An extensive list of smart gadgets, such as baby monitors and security cameras as well as door locks, smart TVs, and power outlets as well as smart toilets, has security holes that might be exploited, according to CNN's reportage. Smart meters, like other connected devices, are vulnerable to hacking and may be taken over by unauthorized parties. Microcontrollers and powerful embedded operating systems (OS) [5] are the most often utilized components in today's smart meters.

The authors briefly explain the need for intelligent CPS security for the Industry 4.0 era in this chapter. The key contributions to this chapter are state-of-art–related studies, CPS 5C level architecture, CPS security issues, and challenges with the possible measures.

The chapter is framed in this way: Section 1.1.1 states the literature review, followed by CPS 5C level architecture in Section 1.3. CPS security issues and challenges are mentioned in Section 1.2. Section 1.3 states CPS security attacks and possible measures. Section 1.4 explains the CPS security issues remedies. Finally, the conclusion is provided in Section 1.4.1.

1.2 Literature Review

In this chapter, the authors have put together a comprehensive report on Industry 4.0 oriented intelligent CPS security based on the studies that have already been done.

The need for current CPS security research and the CPS's unique properties are highlighted by Kim et al. [1] in their discussion of security and the CPS. CPS security challenges and issues are also discussed in depth, as are the numerous CPS security considerations and the variety of CPS assaults that may be uncovered via research into security solutions, attacks, and threats. In addition, research is carried out by authors in order to lead initiatives, explore solutions based on dangers, and provide suggestions for open security concerns.

Using threat analysis, Lu et al. [2] developed a security paradigm for CPS that considers risk assessment from four perspectives: damage, vulnerability, threat, and assets. Emulab, programmable logic controller (PLC), and Matlab are used in this research to simulate CPS. It is possible to use a mix of real-world and computer-generated components in CPS simulations. To make the cyber parts of networked industrial control systems, they came up with a way to use simulations of physical parts and an emulation testbed called Emulab. Matlab Simulink is used to create models of the physical systems, from which C code is created using Matlab. Emulation testbed components may communicate with the RT code that has been developed in the emulation testbed. A PLC may serve as a wonderful example of how people engage with one another. When it comes to actual functioning, PLCs take data from the physical layer, devise a "local actuation strategy," and then transmit orders to the actuators. Supervisory control and data acquisition (SCADA) servers (Masters) provide orders to the PLCs, which they then carry out, and they may also offer comprehensive physical layer data upon request.

The 5C level framework/architecture is explained by Lee et al. [3]. In order to create a CPS application, it is necessary to collect data from machines and the components that make up those devices. Sensors may collect data directly, or data may be collected through controllers or enterprise production systems like ERP, MES, SCM, and CMM. At this stage, two key variables must be addressed. Considering the variety of data, a smooth and tether-free technique to control the data gathering process and transmit data to the central server is necessary where specialized protocols like MTConnect are effective. The second key aspect for the first level, on the other hand, is choosing appropriate sensors (specification and type).

Huang et al. [4] provide a flexible and resilient cybersecurity protection architecture for the suggested industrial CPSs, which may be used to build industrial CPS cybersecurity policies. The most crucial layer in the design is the RT control layer security, because controllers are directly linked to physical devices and their actions have an instantaneous influence on the physical layer. The closed-loop defense framework of intrusion tolerance prevention is designed to ensure the security of the control layer, which takes into consideration the domain characteristics of the industrial control system. A networked water level management system is used as a case study to assess the effectiveness of the cybersecurity protection architecture. For the networked water level control system, an AADL (ocarina architecture analysis and design language) tool suite has been built to verify the process of cybersecurity protection.

CPS are examined by Wurm et al. [5] from a cross-layer viewpoint, with security in various levels being taken into consideration. They have explored the present CPS structure's safety concerns in great depth for the most part. Each layer's security requirements were illustrated in detail by the writers throughout the debate. Authors, instead of considering

CPS as a single entity and attempting to design security solutions for the complete system, identify the many security issues that each layer has and detail countermeasures. To be more specific, this research established three distinct levels of security, covering everything from smart home devices to low-level hardware security.

According to Oliveira et al. [6], OS should grow to become more involved with the underlying hardware and play a more active role in ensuring the safety of their extensions. Computer systems, like the immune system, should regulate the conversations between the original kernel and extensions in order to reduce the risk of security lapses occurring. Toward OS defense, we propose a hardware–software collaborative architecture (HWSW). As a general principle, the OS is designed to provide computers and other devices with the necessary intelligence to enact security rules that safeguard the kernel and its extensions while allowing them to coexist peacefully. As a starting point, the following security rules are considered: As a first step, kernel extensions are not allowed to directly write into the data segments and kernel code, including a restricted section of the stack segment. For the most part, kernel extensions should only communicate with the other extensions and kernels using exposed functions.

Physical unclonable functions (PUF) are a type of semi-conductor-based technology that may be used to create safe connections between cyber and physical substrates based on intrinsic material. The authors of this research [7] discuss how PUF technology might be used to improve security at CPS. With the use of a compositional method, they demonstrate the advantages of merging numerous PUF parts into one random and powerful system-level PUF.

Differential powerful analysis (DPA), simple power analysis (SPA), and other similar approaches are described by Kocher et al. [8]. Using measurements of the power consumption (or other side channels) of a target device, these attacks retrieve secret keys. All main algorithms may be defeated using these techniques. In addition, the authors provide attacks based on power traces and their attributes, which they introduce with an overview. The next assault is a simple DPA attack. They are practical, non-intrusive, and very effective, even in systems with high levels of complexity and noise, where cryptographic calculations account for just a tiny percentage of the total power usage. DPA attacks and developing cryptosystems that are safe even if implemented in hardware that leaks are also discussed by the authors.

Gupta et al. [9] propose a study on smart contract (SC) privacy flaws in software code that may be readily exploited by a malevolent user or threaten the whole Blockchain (BC) ecosystem. The fundamental goal of this research is to look at the function of artificial intelligence (AI) in SC applications and how it may help with concerns like security, privacy, and computing. The authors give a thorough and complete evaluation of SC privacy-preserving and vulnerabilities AI approaches, which is one of the article's major contributions. Second, they have highlighted the newly suggested decentralized AI platforms for SCs in a methodical manner. Finally, the outstanding difficulties and challenges of AI-based SC are discussed. Finally, they presented a retail marketing test case that employs SC and AI to maintain privacy and security.

The necessity of cybersecurity for power systems is described by Stefanov et al. [10]. A testbed framework is a precise and effective tool for identifying security enhancements, CPS vulnerabilities, cyberattack mitigation, and impact analysis. The testbed is used to simulate scenarios of cyberattacks and intrusions on the power system. The IEEE 39-bus system is used in the experiments. The issue of quickly recovering from a blackout caused by cyber assaults has been recognized. The authors also address a SCADA testbed design, cyber intrusion scenarios and effect analyses, and simulation findings.

TABLE 1.1

CPS Security Solutions, Explanation, and Sources

Solutions Type	Explanation	Sources
CPS open issues	Threats, challenges, solutions, and issues for cyber-physical system security for IoT: A survey	Kim et al. [1]
Physical unclonable functions (PUFs), firmware diversity, and machine learning approaches.	A cross-layer approach to cyber-physical system security	Jacob Wurm [2]
CPS simulation using Matlab, PLC, and Emulab.	Cyber-physical system security in a new multilevel framework	Tianbo Lu [3]
A multilayered cybersecurity protection architecture for industrial CPSs that is adaptable and robust.	Securing industrial processes using a cyber-physical system	Shuang Huang [4]
Operating system (OS) safety using the HWSW platform.	CPU and systems benchmarks are estimated using Ianus' performance with low overhead. Describes a hardware–software architecture to provide OS security.	Oliveira et al. [6]
PUF	Digital fingerprints may be used in semiconductor devices like microprocessors as a unique identifier.	Al Ibrahim et al. [7]
DPA	To prevent attack players, research the entrance to the plain text and cipher text information	Kocher et al. [8]
Improve SCADA systems' cybersecurity	Analyzing the impact of a CPS and its security	Stefanov et al. [10]

CPS simulation and modeling are the two important considerations in terms of system development. Oks et al. [11] suggest the idea to develop CPS with variable size, modularity, and scalable complexity. The authors also proposed the industrial CPS reference architecture for designing, which enables industrial CPS representation in various configurations regarding scaling, complexity, and representation.

For cloud-based CPS (C2PS), this work [12] provides an analytical description of a digital twin architectural reference model in which every real-world entity is associated with a cloud-hosted cyber thing. It is possible to build peer-to-peer (P2P) links either through direct physical contact or via indirect digital twin connections in the cloud. Computation, control, and communication are all analytically modeled in this article. On the basis of the suggested cloud-based CPS reference model, they also give the design specifications for an automated car driving assistance application. A telematics-based car driving assistance application is also described in depth in the C2PS architecture presentation (Table 1.1).

1.3 CPS 5C Level Architecture

The proposed five-level CPS structure, specifically the 5C architecture, presents a gradual framework for rising and organizing a CPS for advanced functions. A CPS, in general, consists of two major useful elements [1]: superior connectivity, which ensures RT information attainment from the physical world and data advice from the cyber room; and intelligent

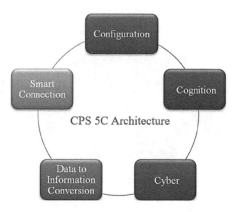

FIGURE 1.2
CPS 5C level architecture.

data organization, analytics, and computational ability, which construct cyberspace [2]. On the other hand, such an obligation is very theoretical and not sufficient for execution purposes in general. In comparison, the 5C architecture [3] presented here obviously describes, in a chronological workflow way, how to build a CPS from the early data attainment, to analytics, to the final value formation. The 5C architecture is shown in Figure 1.2.

Ordinary illustrations of CPSs consist of industrial control systems (now particularly interesting in light of the so-called Industry 4.0 developments, which are in essence a tight integration of information systems, company data, and computerized production systems), mechanized vehicle and aircraft controls, wireless sensor networks, smart grids, and almost every device characteristically included in the IoT. Again, numerous medical devices, mainly implantable ones such as pacemakers, are also becoming CPSs. Such systems are frequently key elements of contemporary serious infrastructure and therefore are necessary to our society's financial feasibility and social solidity. For these reasons, the security, safety, and reliability of CPS are very crucial for a sustainable society [3].

1.4 CPS Security Issues and Challenges

As the importance of cybersecurity develops, it is critical that CPS security notices be sent more quickly. Communication security, physical security, and operational and control security are the three main facets of CPS security. Security in the physical environment focuses on data defense in a networking atmosphere, data aggregating in loosely connected systems, giving out, and large-scale dividing; security in communication focuses on information defense and control scheme responsibilities in the face of cyberattacks [2]. Smart devices' security risks may be divided into six groups depending on the methods used to attack them. Figure 1.3 shows the seeds of smart device security vulnerabilities.

1.4.1 Security Framework

Now examining the safety of CPS is centered on the protection of information and scheming. Information security solves the troubles of data compilation, processing, and distribution non-destructively in a large-scale, high-mix, joint/independent system environment.

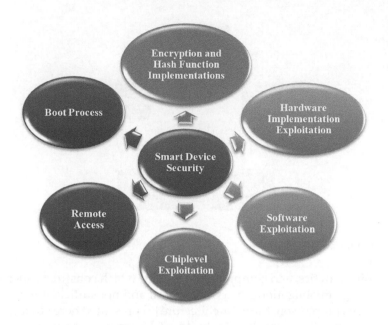

FIGURE 1.3
Security vulnerabilities in smart devices.

Its main points are enhancing obtainable safety methods, user solitude protection, well-organized processing of large data encryption, etc. Control security solves the problem of planning in networked systems with open connections and planning that is not very tightly linked. It focuses on regaining power through attacks on system opinion and control algorithms [2].

1.4.1.1 Perception Issues and Challenges

In a range of perceptions of its surroundings, the CPS includes sensors, actuators, and physical ecological issues. The physical atmosphere takes out the control unit's training, and extensive and supervised data are broadcast to the control unit.

Real-time digital simulations (RTDS) were primarily used in the power industry to generate field machine output and input signals. In the hydroelectric system, the field machines were built with pumps, sensors, and tanks. The substantial atmosphere has points with admission to the apparatus, and illegal things can enter and influence the network. A person must first execute the program in order to protect the system from harmful hackers. It does not ensure the execution of any additional instruments by the program, except that the basic binary code can be believed. Here, IoT safety concerns comprise apparatus and front-facing sensors, internal IT procedures, and internetworks, whereas IoT confidentiality regards comprise machine confidentiality, confidentiality during transmission, and private data repository space issues [1].

1.4.1.2 Challenges and Issues in the Application Layer

It is the responsibility of the CPS application to control, measure, and monitor the physical environment. Real-world devices and systems are being used in research to build the control. There is a wide range of security considerations in today's application landscape because of the diversity of the applications being used, including smart industries, smart cities, smart buildings, smart grid, and smart healthcare [1].

1.4.1.3 Communication Issues and Challenges

Using the different forms of data (such as events, control status, and measurements) received from the control state, the CPS's communication layer issues directives to the appliances and systems liable for the web connecting the control system and devices. Instead of focusing on the application layer, communication needs influence the physical, middle, and networking layers. Each device's radio connection qualities toward various devices, like the router, the overall network load and capability, and other factors all play a role in the CPS communication area difficulties that must be addressed before picking a suitable communication standard. When dealing with CPS security challenges, it is important to find a balance between the security measures that are put in place at different levels and the cost of putting them into action [13]. As a transmission difficulty, SDN (software-defined networking) may alter the rules of the routing, and changes in the CPS context affect routing control and quality of service, maintaining RT needs, dependability, and security at the same time [1].

1.5 CPS Security Attacks and Possible Measures

In the CPS, individual tiers may be physically or aggressively disrupted, causing substantial physical harm. Cyberattacks against CPS sensors and networks are also possible [14]. Issues to the perceptual layer include cyberattacks on actuators and sensors, mainly on IoT; challenges to the communication tier containing system communications, data corruption, and threats; and dangers to the application tier include malicious code and counterfeit attacks. These different dangers are grouped as CPS security issues (Figure 1.4). Issues to CPS security include ways to secure physical, cyber, and system space. These solutions demonstrate the importance of smart surroundings like smart grids, smart cities, smart healthcare, smart industries, and smart buildings.

1. **Perception attacks**: The memory functions and physical environment, such as sensors' RFID make up the perceptual layer. These devices are often housed outside, exposing them to physical attacks when devices or components are replaced [15]. Also, most CPS security study concentrates on cyberspace, neglecting the vulnerability of the hardware layer, necessitating study on that topic. Node reputation, data password threats, data tampering, perceptual data corruption, denial of service attacks, fault attacks, electromagnetic interference, electromagnetic leaks, line failures, and equipment failures are all physical risks.

   Node reputation: Attacks against the reputation of nodes include the capture of nodes, the use of phantom nodes, and outages of nodes. To ensure the safety of the whole system, node capture obtains or surrenders encryption key information on behalf of the nodes [16]. By creating a new node in the network, the false node is able to send out harmful data. Node energy may potentially be used to attack data integrity and execute DoS attacks [17]. By interrupting the node's capacity to receive and gather data, an attack known as a "node abort"

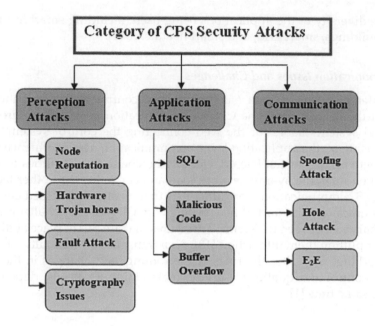

FIGURE 1.4
Category of CPS security attacks.

compromises integrity and availability [18]. Stability, integrity, availability, and confidentiality are the primary goals of these diverse reputation attacks.

Hardware trojan horse: The behavioral, activation, and physical aspects that go into the Troy categorization system are what set it apart. It is possible to determine how many components have been inserted, erased, or damaged by looking at the chip's physical attributes. Troy's destruction and activation functions are governed by the activation attribute. Trojan horses are known for their destructive activities because of their behavioral features [19].

Fault attack: The data password is restored and the internal circuitry is reverse engineered using a fault attack, which involves fault actions on the target device that are purposely induced. All systems' critical paths confront circuit control issues when a fault attack occurs, as is well-known. In addition, a fault attack may use many approaches that may be classified as either local or global [20].

Cryptography issues: These distributed hardware devices gather, transmit, process, and store CPS security data, which are intentionally attacked by hardware or cyber layers. Hardware-based solutions may potentially attack the endpoint in an unsupervised situation [21]. CPS terminals with low resources have increasingly relied on lightweight cryptography. DPA attacks target the AES SubBytes and AddRoundKey outputs when a device is handling cryptographic secrets [22].

2. **Application attacks**: This layer is where a huge amount of users' information is collected, and it may lead to hacks that result in information loss, the theft of individual data like health and habits, and the illegal usage of the device itself. Vulnerability attacks, DoS/DDoS, control command forgery and database attacks,

malicious code, unauthorized access, and user privacy disclosure are all examples of attacks on the application layer.

Structured query language (SQL): For content manipulation and structural modifications, most corporate database systems may be accessed through SQL commands. It is possible to enter SQL statements into a query that are not intended by the online application, and this is known as a SQL injection attack. As a result, one can have the ability to perform a wide range of operations on the database. A cyberattack on the SCADA system using SQL injection has been shown to have a significant influence on its security in current data and online accessibility.

Malicious code: Malicious code may harm a network by attacking a user's application through malicious code from worms and viruses. An injected payload might cause an exploited method to execute or crash a harmful program, depending on the payload. An attack's capacity to conceal specifics without invalidating the stages in between cannot be overstated.

Buffer overflow: When a software's intended behavior is compromised by a buffer overflow attack, the program is unable to perform its regular functions. Function-pointer manipulation and stack smashing are two common attacks. Passwords are rebuilt, the material is altered, and malicious code is executed as a consequence of this kind of attack [23].

3. **Communication attacks**: The CPS's communication must be safeguarded in order to prevent any flaws in the system. As many servers, infrastructure devices, and field devices as possible should be protected by a CPS that can scale up to meet their needs in terms of lightness, resilience, and scope. Transmission hazards include routing loop, hole attacks (e.g., gray hole attacks, black hole attacks, wormhole attacks, sinkhole attacks), Sybil attacks, trap doors, flood attacks, control attacks, DoS attacks, and routing attacks. These are only a few examples of the many threats that data transmissions are subject to. There are several dangers to be aware of.

Spoofing attack: As the name suggests, this kind of attack involves sending ICMP (internet control message protocol) messages to the target network repeatedly with the same destination address as one of the computers on that network. A PLC may send a harmful order to an actuator if it receives a changed message from a SCADA system. An ARP spoofing meson fools the opposite party's data packet into believing it is a gateway by using an address determination protocol message.

Hole attack: Gray hole attacks, black hole attacks, and sinkhole attacks are all examples of hole attacks. An attack on the WSN routing protocol known as a "wormhole" involves two types of assaults. Using wormhole tunnels, two suspicious elements search for data packets in one area of the network and transfer them to the remainder of the network via them, providing a false sense of closeness. Intruders may be able to take over many pathways in the network, causing the routing system to malfunction [24]. The sinkhole attack is the suspicious element of the sinkhole that provides the most direct path to the base station. Data may be tampered with before being deleted or sent to the base station by sinkhole nodes [25].

End-to-end (E2E): DDoS and DoS threats, encryption techniques, and key management may all be made more difficult using E2E. Furthermore, in large-scale systems, CPS devices are scattered throughout a wide geographic area. Secure communication is needed for dispersed appliances, which may be revealed to other stakeholders and pose economic and safety threats. As a result, a large-scale CPS should be used to secure the E2E side of the network against cyberattacks. Security support for the safeguard of individual data in transmissions for the gathering of data from energy sources or smart meters is also designated as a non-negotiable kind of E2E security assistance. Large CPSs might lose reliability without E2E security support [26].

1.6 CPS Security Issues Remedies

Application privacy [27, 28], network security, and data protection are all part of CPS [29, 30] security's scope, which extends from the physical to the application environments. As illustrated in Figure 1.5, the CPS security solutions may be divided into four main categories: application solutions, malicious code detection, network access detection, and device protection.

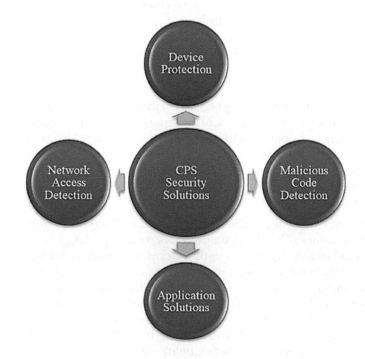

FIGURE 1.5
CPS privacy and security remedies.

Application solution: Smart grids, healthcare, industries, cities, buildings, and other connected remedies are among the many smart settings for which application solutions are appropriate [31–35].

Malicious code detection: On the internet, many malware detection algorithms have been presented [36]. However, because these technologies only identify malicious code on the internet, they cannot be used to control dangerous programs spread in the CPS.

Detection of the network access: A wormhole and defense attack detection technique that uses the control system detection and SDN is included in the system and network access defense (SDN) [37–39]. However, in this area, rigorous research is carried out globally.

Device protection: With this in mind, it is important to create a secure hardware platform that is resistant to cyberattacks. Different researchers have proposed various architectures to combat the CPS attacks. That is, the system must also display the ideal characteristics of a CPS: availability, privacy, reliable confidentiality, efficiency, and security.

1.7 Conclusion

Because CPS security is not a so much explored domain that is distinct from the present communication scenario, there has been little research in this area. Applications, processes, data, and sensors are all sent over the CPS, which is a transmission medium that can handle a wide range of information exchanges. In this study, the numerous security ideas, risks, and cures confronting the CPS are classified, and a solution to each threat is presented. In the context of CPS security, there were a wide range of concerns and challenges, as well as a wide range of CPS-related surveys. As a result of the attacks, the CPS security review examined the available solutions as well as possible future paths for analysis. As part of our investigation of CPS security, we examined the link between assaults and possible solutions. Because the CPS is being used in so many different smart contexts, including the smart grid, healthcare, industry, and cities, the security of the CPS should be a top priority at all times. Because smart environments are becoming more common, CPS security is projected to grow in importance.

References

1. Kim, N. Y., Rathore, S., Ryu, J. H., Park, J. H., & Park, J. H. (2018). A survey on cyber physical system security for IoT: Issues, challenges, threats, solutions. *Journal of Information Processing Systems*, 14(6), 1361–1384.
2. Lu, T., Xu, B., Guo, X., Zhao, L., & Xie, F. (2013, March). A new multilevel framework for cyber-physical system security. In *First international Workshop on the Swarm at the Edge of the Cloud*.
3. Lee, J., Bagheri, B., & Kao, H. A. (2015). A cyber-physical systems architecture for industry 4.0-based manufacturing systems. *Manufacturing Letters*, 3, 18–23.

4. Huang, S., Zhou, C. J., Yang, S. H., & Qin, Y. Q. (2015). Cyber-physical system security for networked industrial processes. *International Journal of Automation and Computing, 12*(6), 567–578.

5. Wurm, J., Jin, Y., Liu, Y., Hu, S., Heffner, K., Rahman, F., & Tehranipoor, M. (2016). Introduction to cyber-physical system security: A cross-layer perspective. *IEEE Transactions on Multi-Scale Computing Systems, 3*(3), 215–227.

6. Oliveira, D., Wetzel, N., Bucci, M., Navarro, J., Sullivan, D., & Jin, Y. (2014). Hardware-software collaboration for secure coexistence with kernel extensions. *ACM SIGAPP Applied Computing Review, 14*(3), 22–35.

7. Al Ibrahim, O., & Nair, S. (2011, July). Cyber-physical security using system-level PUFs. In *2011 7th International Wireless Communications and Mobile Computing Conference* (pp. 1672–1676). IEEE.

8. Kocher, P., Jaffe, J., Jun, B., & Rohatgi, P. (2011). Introduction to differential power analysis. *Journal of Cryptographic Engineering, 1*(1), 5–27.

9. Gupta, R., Tanwar, S., Al-Turjman, F., Italiya, P., Nauman, A., & Kim, S. W. (2020). Smart contract privacy protection using AI in cyber-physical systems: Tools, techniques and challenges. *IEEE Access, 8*, 24746–24772.

10. Stefanov, A., & Liu, C. C. (2014). Cyber-physical system security and impact analysis. *IFAC Proceedings Volumes, 47*(3), 11238–11243.

11. Oks, S. J., Jalowski, M., Fritzsche, A., & Möslein, K. M. (2019). Cyber-physical modeling and simulation: A reference architecture for designing demonstrators for industrial cyber-physical systems. *Procedia CIRP, 84*, 257–264.

12. Alam, K. M., & El Saddik, A. (2017). C2PS: A digital twin architecture reference model for the cloud-based cyber-physical systems. *IEEE Access, 5*, 2050–2062.

13. Burg, A., Chattopadhyay, A., & Lam, K. Y. (2017). Wireless communication and security issues for cyber–physical systems and the Internet-of-Things. *Proceedings of the IEEE, 106*(1), 38–60.

14. Peng, Y., Lu, T., Liu, J., Gao, Y., Guo, X., & Xie, F. (2013, October). Cyber-physical system risk assessment. In *2013 Ninth International Conference on Intelligent Information Hiding and Multimedia Signal Processing* (pp. 442–447). IEEE.

15. Shafi, Q. (2012, June). Cyber physical systems security: A brief survey. In *2012 12th International Conference on Computational Science and Its Applications* (pp. 146–150). IEEE.

16. Mahmoud, R., Yousuf, T., Aloul, F., & Zualkernan, I. (2015, December). Internet of things (IoT) security: Current status, challenges and prospective measures. In *2015 10th International Conference for Internet Technology and Secured Transactions (ICITST)* (pp. 336–341). IEEE.

17. Zhao, K., & Ge, L. (2013, December). A survey on the internet of things security. In *2013 Ninth International Conference on Computational Intelligence and Security* (pp. 663–667). IEEE.

18. Bhattacharya, R. (2013). A comparative study of physical attacks on wireless sensor networks. *International Journal of Research in Engineering and Technology, 2*(1), 72–74.

19. Tehranipoor, M., & Koushanfar, F. (2016). A survey of hardware trojan taxonomy and detection. *IEEE Design & Test of Computers, 27*(1), 10–25.

20. Khelil, F., Hamdi, M., Guilley, S., Danger, J. L., & Selmane, N. (2008, November). Fault analysis attack on an FPGA AES implementation. In *2008 New Technologies, Mobility and Security* (pp. 1–5). IEEE.

21. He, W., Breier, J., Bhasin, S., & Chattopadhyay, A. (2016, May). Bypassing parity protected cryptography using laser fault injection in cyber-physical system. In *Proceedings of the 2nd ACM International Workshop on Cyber-Physical System Security* (pp. 15–21).

22. Kocher, P., Jaffe, J., Jun, B., & Rohatgi, P. (2011). Introduction to differential power analysis. *Journal of Cryptographic Engineering, 1*(1), 5–27.

23. Zhu, B., Joseph, A., & Sastry, S. (2011, October). A taxonomy of cyber attacks on SCADA systems. In *2011 International Conference on Internet of Things and 4th International Conference on Cyber, Physical and Social Computing* (pp. 380–388). IEEE.

24. Gupta, G. (2015). Frequency based detection algorithm of wormhole attack in WSNs. *International Journal of Advanced Research in Computer Engineering & Technology, 4*(7), 3057–3060.

25. Pirzada, A. A., & McDonald, C. (2005, May). Circumventing sinkholes and wormholes in wireless sensor networks. In *IWWAN'05: Proceedings of International Workshop on Wireless Ad-hoc Networks* (Vol. 71).
26. Kim, Y., Kolesnikov, V., & Thottan, M. (2016). Resilient end-to-end message protection for cyber-physical system communications. *IEEE Transactions on Smart Grid, 9*(4), 2478–2487.
27. Banerjee, J., Maiti, S., Chakraborty, S., Dutta, S., Chakraborty, A., & Banerjee, J. S. (2019, March). Impact of machine learning in various network security applications. In *2019 3rd International Conference on Computing Methodologies and Communication (ICCMC)* (pp. 276–281). *IEEE.*
28. Chakraborty, A., Banerjee, J. S., & Chattopadhyay, A. (2020). Malicious node restricted quantized data fusion scheme for trustworthy spectrum sensing in cognitive radio networks. *Journal of Mechanics of Continua and Mathematical Sciences, 15*(1), 39–56.
29. Das, K., & Banerjee, J. S., (2022). Green IoT for intelligent cyber-physical systems in industry 4.0: A review of enabling technologies, and solutions. In *GCAIA 21: Proceedings of 2nd Global Conference on Artificial Intelligence and Applications* (pp. 463–478).
30. Das, K., & Banerjee, J. S., (2022). Cognitive radio-enabled internet of things (CR-IoT): An integrated approach towards smarter world. In *GCAIA 21: Proceedings of 2nd Global Conference on Artificial Intelligence and Applications* (pp. 541–555).
31. Pandey, I., Dutta, H. S., & Banerjee, J. S. (2019, March). WBAN: A smart approach to next generation e-healthcare system. In *2019 3rd International Conference on Computing Methodologies and Communication (ICCMC)* (pp. 344–349). *IEEE.*
32. Biswas, S., Sharma, L. K., Ranjan, R., Saha, S., Chakraborty, A., & Banerjee, J. S. (2021). Smart farming and water saving-based intelligent irrigation system implementation using the internet of things. In Bhattacharyya, Siddhartha, Dutta, Paramartha, Samanta, Debabrata, Mukherjee, Anirban, & Pan, Indrajit (Eds.) *Recent Trends in Computational Intelligence Enabled Research* (pp. 339–354). Academic Press.
33. Roy, R., Dutta, S., Biswas, S., & Banerjee, J. S. (2020). Android things: A comprehensive solution from things to smart display and speaker. In *Proceedings of International Conference on IoT Inclusive Life (ICIIL 2019), NITTTR Chandigarh, India* (pp. 339–352). Springer, Singapore.
34. Chakraborty, A., Singh, B., Sau, A., Sanyal, D., Sarkar, B., Basu, S., & Banerjee, J. S. (2022). Intelligent vehicle accident detection and smart rescue system. In Mandal, Jyotsna Kumar, Misra, Sanjay, Banerjee, Jyoti Sekhar, Nayak, Somen (Eds.) *Applications of Machine Intelligence in Engineering* (pp. 565–576). CRC Press.
35. Chattopadhyay, J., Kundu, S., Chakraborty, A., & Banerjee, J. S. (2018, November). Facial expression recognition for human computer interaction. In *International Conference on Computational Vision and Bio Inspired Computing* (pp. 1181–1192). *Springer, Cham.*
36. Rathore, S., Sharma, P. K., Loia, V., Jeong, Y. S., & Park, J. H. (2017). Social network security: Issues, challenges, threats, and solutions. *Information sciences, 421,* 43–69.
37. Banerjee, J. S., Chakraborty, A., & Chattopadhyay, A. (2021). A decision model for selecting best reliable relay queue for cooperative relaying in cooperative cognitive radio networks: The extent analysis based fuzzy AHP solution. *Wireless Networks, 27*(4), 2909–2930.
38. Chakraborty, A., Banerjee, J. S., & Chattopadhyay, A. (2019). Non-uniform quantized data fusion rule for data rate saving and reducing control channel overhead for cooperative spectrum sensing in cognitive radio networks. *Wireless Personal Communications, 104*(2), 837–851.
39. Banerjee, J. S., Chakraborty, A., & Chattopadhyay, A. (2022). A cooperative strategy for trustworthy relay selection in CR network: A game-theoretic solution. *Wireless Personal Communications, 122*(1), 41–67.

2

Issues in Authentication in Augmented Reality Systems

Aman Srivastava, Pranjali Shah, and Ankita Agarwal
Babu Banarasi Das Institute of Technology and Management, Lucknow, India

Hemlata Pant
Maharishi University of Information Technology, Lucknow, India

CONTENTS

DOI: 10.1201/9781003241348-2

2.1 Introduction

The fusion of the digital world's components and an individual's perception of the real-world forms the augmented reality (AR) [1–7]. It alters the way one sees the real world and thus forms an interface between fantasy and reality. It digitally manipulates the information about the real-world environments of the user by making it interactive. Thus, AR amalgamates real and virtual world combinations along with interaction. So, AR is something which is itself not real but supplements the reality that exists beforehand. It is not the replacement of the digital world with the real world, but instead it is the placement of objects of the digital world into the real world [8]. The use of AR is not only confined to games and entertainment but it is expanded to businesses as well. AR can be exhibited on devices like screens, mobile phones, and glasses. Not all but many modern devices support AR. Smartphones, tablets, smart glasses, and smart lenses fall under the category of devices that are suitable for AR.

The enhancement of digital communication through AR has also brought an increase in several digital threats, thus emphasizing on the critical need of security and protection in AR technology. The possible security concerns posed by AR needs to be addressed before its widespread use. Such technologies are used in shared spaces and thus posing new security challenges which also includes challenges in authentication AR systems not only present challenges but also the opportunities to improve security and privacy. Authentication is the basic process of identifying the original owner of the device. Because the AR technologies are so new, there is not a single mechanism that is completely reliable and secured. As the security threats and attacks are being launched every day, researchers have introduced different kinds of authentication mechanisms for AR devices. So, in this chapter,

  i. We have addressed most of the commonly used mechanism that are available today
 ii. We have studied these protocols and provided brief information
iii. Compared them on the basis of certain performance and security measures

which led to create a survey table that can provide a clear understanding and overview of security of these authentication mechanisms at a glance. We have differentiated the authentication mechanisms on the basis of the following criteria:

  i. Cryptography or Algorithm used
 ii. Security against several potential attacks

This chapter is organized into different sections as follows: Section 2.2 provides the background and motivation of the study and Section 2.3 is about the history of AR systems. Section 2.4 consists of comparison of AR with virtual reality and Section 2.5 contains several applications of AR in different fields. Section 2.6 highlights the integration of AR with CPS. Section 2.7 briefly summarizes the overview and information about several authentication protocols used in AR. Section 2.8 discusses the various factors on which the protocol is evaluated and consists of comparison study. And finally, the chapter concludes with Section 2.9.

2.2 Background and Motivation

Several studies have been carried out by researchers to minimize the security issue and build a better authentication protocol. Most of the protocols mentioned below are secured in one way or another, but most of them are not completely secured and reliable for widespread usage of AR devices. To the best of our knowledge, no prior study on these protocols have been done till date, we are the first to compare these protocols and address the security issues with them. With the goal to build a reliable and secured protocol one day, this chapter provides a comprehensive study on existing authentication protocols in AR systems.

Gutmann et al. [9] have explored the human-based computation scheme, and they have used the private channels for devices to implement memorization of an authentication secret. Although this protocol ensures reliability to some extent because of the usage of semantics, human-based computing ensures the resistance against untrusted device, but in case somebody knows the answers to those logical questions, it is vulnerable to impersonation and denial-of-service (DoS) can also be carried on the channels.

Krishna et al. [10] have implemented the usage of electroencephalogram (EEG) signals along with eye-tracking mechanism. They used machine learning approach to process the EEG signals and support vector machines (SVM) to construct a 64×1 vector. Then, the eye-tracking signals were used to train the random forest classifier with 100 trees. Although this seems to be using more sophisticated features with machine learning algorithms, but chance of man-in-the-middle (MiTM) attack and DoS happening is a nonzero factor.

Gaebel et al. [11] have employed the scheme using Diffie–Hellman exchange cryptographic algorithm to establish the shared-symmetric key and facial recognition to the final authenticity verification. This protocol does ensure security against MiTM attacks and reduced user-device authentication, and facial recognition provides extra security measure against impersonation.

Wang et al. [12] have proposed the Nod-to-Auth protocol that uses IMU sensors to unlock the headsets with nodding gestures. They have used machine learning model with features like changes in neck length, average neck length, and head orientation during head gestures, etc. This protocol is inexpensive and simple to implement in AR/virtual reality (VR) devices as most of the headsets are now equipped with IMU sensors. If an adversary can imitate the head gestures of the user, then they can unlock the device. Also, security against MiTM and DoS is present.

Bhalla et al. [13] have used sensors present in the headset to get the spatial and behavioral patterns of the user and then with the machine learning algorithms like k-NN, SVM, and random forest, different parameters are evaluated. Wazir et al. [14] have proposed the usage of doodle for authentication. They have used AR Drawing (Google Creative lab) along with smartphone screen to generate doodle in the AR space and then coordinates of the doodles are matched and authentication is successful.

More detailed definition and review of several authentication mechanisms are discussed in Section 2.6.

2.3 History of AR

AR has seen tremendous growth over the last few years but it has been around for an age. Early part of the 20th century saw the first reference to AR. Recent developments made are evidence of the fact that AR is expected to be the next big asset commercially. The primary sort of AR device was designed and developed by a professor Ivan Sutherland who was from Harvard University. He did this along with his student Bob Sproull. He named this first AR device as The Sword of Damocles, which was basically a head-mounted device and it was made to hang from the ceiling. It gave users the feel of alternate reality because of experiencing computer graphics. This concept was closer to VR [15].

The next development in this domain was made in 1974 by Myron Krueger which was called Videoplace [8]. It produced shadows on the screen with the help of the combination of projection system and video cameras. This gave the user a feeling of interactive environment. The term augmented reality was coined by a researcher from Boeing, named Tom Caudell, in 1990. The first real and operational AR system was created by Louis Rosenburg, in 1992, who belonged to the USAF Armstrong's Research Lab. It was named virtual fixtures. This could be said as the previous version of what is currently done today by most of the AR systems [16].

After 2 years, in 1994, a theater production was created which used AR. "Dancing in Cyberspace" was the name of the theater and it was produced by Julie Martin. It presented the dance of the acrobats, in and around the virtual objects.

The major progression in this domain was seen in 2000, when a software called ARToolKit was created and released by Hirokazu Kato, who was from the Nara Institute of Science and Technology in Japan. Capturing of real-world actions and then merging it with interactions of virtual objects could be done with the help of this software [17]. In the year 2009, with the help of this software, AR was made available to Internet Browsers.

AR has made a lot of progress since its origination, and the evolution it has made in the past few years has proved to be more promising.

2.4 Comparison with VR

Digitally, both of these technologies, AR and VR provide an opulent experience. On one hand, VR consists of 75% real content, whereas 25% of the content is virtual. On the other hand, AR consists of 25% virtual content and the rest of the 75% is real. VR necessitates the use of headphones while no such need is there with AR. VR controls one's visual senses, whereas when using AR, the presence in the real environment is always felt. Although AR is one in which something is added to the things which already exist, VR is entirely immersive as it generates a three-dimensional image. As a result, in VR, it is felt as if the user is surrounded by the image [17]. For finest quality experiences, AR requires higher bandwidth, more than 100 mbps, whereas VR needs lower bandwidth of at least 25 mbps to stream. More advanced and sophisticated technology is required for VR in comparison to AR. AR systems are associated with the need of sensors to collect the data from the real world but there is no such need with VR as it isolates the user from the real world. Implementation of AR is less expensive than the implementation of VR as it requires high-cost equipment.

AR products like Google glasses are available in comparison to VR products as they are yet not available as it needs something to immerse the user into the different world entirely. The algorithms and software used by VR are more complex than those used by AR. VR also needs more graphics along with processing power. Despite all these differences between AR and VR, both of these are highly promising for the future in different domains of education, gaming, market, and commerce. Both the technologies together upgrade the user experience by combining the virtual and real world.

2.5 Applications

In recent years, a tremendous growth in usage and applications of AR has been witnessed. It can be accredited to the facilities that the consumers get because of AR. They can visualize and imagine using the product and experience the service even before actually buying it. For creation of AR apps by developers, in 2017, Apple launched ARKit and Google launched ARCore for Android. Because of sophistications and cost-effectiveness of AR, the demand and investment in AR will increase in coming times.

2.5.1 Medical

The need of AR in the medical industry can be correlated to the need of the visualization of the patient and medical data in the same physical space. Furthermore, the use of AR in the medical domain can be seen in ultrasound imaging [18–20].

2.5.2 Military

For the exhibition of the scene of real battlefield, AR is used. For providing military training in urban terrain, AR can be used. With the help of AR, an animated terrain could also be displayed which can be helpful in military intervention planning [21].

2.5.3 Robotics

For communicating information to humans, AR technique is used by the robots. This is mainly done for the presentation of complex information. The combination of AR technique and surgical robot system can be used for head surgery [22].

2.5.4 Visualization

For overlaying computer graphics into the real world, AR is used. Through AR, invisible concepts and events can also be visualized. This can be done through the superimposition of virtual objects onto physical environment. Concepts of abstract science can also be visualized through AR [23, 24].

2.5.5 Education

New opportunities in the field of teaching and learning is possible through the implementation of AR. Students can visualize and experience phenomena that is not possible in the real world. Interaction of two- and three-dimensional synthetic objects can also be done with the help of AR [25].

2.5.6 Tourism

Information about the cultural heritage sites and their archaeological information can be provided to visitors through an AR-based project on onsite guide. This was done for the enhancement of experience in cultural tourism [20, 26].

2.5.7 Marketing

Automotive industry is the industry where AR was first used for the advertisement purpose. Three-dimensional models of the cars to be advertised were shown on screen. This approach was then widely spread to other areas like movies, shoes, and furniture. This is very helpful for the users as they can find the appropriate product for themselves by testing the color, accessories, and model [27, 28].

2.5.8 Manufacturing

AR can be used in enhancement of manufacturing processes along with the development of product and process. It can change a person's perception of the surroundings. This will lead to reduced cost and better quality of the manufactured products [29, 30].

2.5.9 Entertainment and Games

AR is widely used for creation of various games. It is also used in sports broadcasting. Through AR, swimming pools and football fields can also be easily prepared [31].

2.6 Integration of AR with CPS

AR system can be used for prototyping the cyber-physical system (CPS) for planning, perception, learning, and control. Omidshafiei et al. [32] have developed a framework for prototyping autonomous vehicles. The assimilation of computerized network systems and physical processes constitutes CPS. The key feature of CPS which enables its future scope of development is its ability to interact with the physical world and its processes. Furthermore, with the help of computation and communication, it can broaden the capabilities of the physical world. Apart from this, the other important feature of CPS is its bi-directional coordination. This coordination is made between virtual world to physical world and from physical world to virtual world.

One way of integrating CPS and AR is integrating CPS in the construction industry. This will lead to improvement in productivity and efficiency in these industries. For the facilitation of this coordination between virtual and physical world, a mechanism needs to be developed. This also involves designing the correct position of placement of sensors on key physical components [33].

The alternative of doing this integration is integrating the concept of CPS with mobile AR. The growth of AR has seen a shift toward handheld systems. This will result in an abstract view of geospatial data and visualization of underground infrastructure. There are huge challenges involved with the development of CPS because of its complexities. So, different issues need to be resolved at different levels for effective integration.

2.7 Authentication in AR

Authentication in AR systems is used to provide a layer of verification to identify whether the user is genuine and claims who he is. As the AR systems and VR systems evolve into standalone devices, it is not simple for users to use authentication mechanisms as they use in their smartphones and/or computers, so there are several authentication protocols that are proposed by different researchers. This section briefly discusses some authentication techniques that are proposed and/or used for AR systems (Figure 2.1).

2.7.1 ZeTA—Zero Trust Authentication

Devices and channels that are used in the process of authentication need to be reliable for proper authentication. ZeTA (zero trust authentication) protocol is an authentication scheme based on human-based computation to reduce the threat of using untrusted devices and channels in the process of authentication. The idea behind ZeTA protocol is the memorization of an authentication secret. When the user enrolls, ZeTA generates and assigns the secret to the user. The enrollment is done with the help of a private channel between the system and the user. This secret consists of different words with logical connection like AND, OR, and NOT between them. Confirmation or denial of a particular attribute on the basis of semantics and established logical connection between the words is done. The protocol tests the user's semantic knowledge of the words and their meanings used in the secret. Personalized questions are asked along with answers that are related

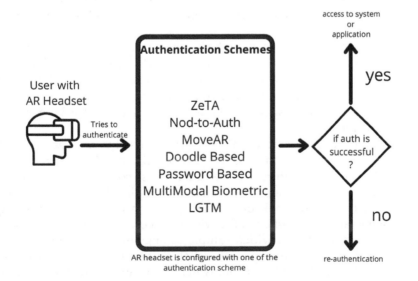

FIGURE 2.1
Authentication flow in augmented reality systems.

with the general knowledge of the secret. The significance of the protocol lies in the fact that it uses human processing. The knowledge which is stored in our semantic memory is associated with a considerable amount of inter-human correlation. The notion behind this protocol does not require the user to conduct any mathematical operation. The user is challenged to confirm or deny the things asked which are related to the semantics and attributes of the secrets. As a consequence, the scaling of ZeTA can be done to arbitrary security levels [9].

This protocol is divided into three categories:

   i. It uses semantics and human-based computing to solve the problem of untrusted devices and channels, with a formal specification and security analysis.
   ii. This protocol presents a security study to evaluate the proposed authentication protocol's expected performance.
   iii. The results of a usability research looks into the possibility of using human computation in this situation.

Resistance against untrusted devices and communications and shoulder-surfers are the two primary properties that the suggested protocol should have. A challenge–response interaction style is proposed to accomplish this. The user's portion must be completed mentally and without the usage of any external artifacts.

2.8 Multimodal Biometric Authentication

EEG signals can be used in conjunction with a wearable headset (i.e. AR/VR) to provide an additional non-intrusive input modality. A fundamental difficulty for brain–computer interface (BCI) algorithms is the poor generalization performance of EEG data. The possibilities of using this technology for facial recognition on smartphones emerge as a result of these user differences.

Furthermore, data from eye tracking, which is generally available in such headsets, is used in the evaluation. Biometric authentication methods are being developed for each of these systems, as well as their fusion. EEG motor imagery and eye-tracking data that are publicly available are used to build evaluation paradigm.

The three main steps in this system are EEG authentication, eye-tracking authentication, and multimodal fusion.

The ERPs generated during a motor imagery task are used to analyze EEG data in EEG authentication. To epoch and pre-process the EEG signals, the MNE program [34] is used. To create a 64 X1 vector of a feature, the maximum cross-correlation value is used. SVM with linear and radial basis function (RBF) kernels are applied to this feature vector. In eye tracking authentication, feature vectors made up of concatenated eye-tracking signals were used to train a random forest classifier with 100 trees. The model predicts an array of posterior probabilities that the provided sample belongs to each of the $n = 5$ authorized users and the unauthorized group of possible labels (total $n + 1 = 6$ bins). Two fusion methods have been used in multimodal fusion: weighted mean and fusion using SVM with linear kernels, both of which produce a normalized match-score from individual predictions [35]. In this procedure, when evaluating on the basis of dataset composition, an SVM with linear kernels and random forest classifiers outperformed the individual modalities in terms of recognition accuracy.

2.8.1 Looks Good To Me

In order to share the content between AR headsets, user authentication plays a very crucial role. Traditional authentication protocols fail to meet the needs in case of unavailability of a central authentication server. Looks Good To Me (LGTM) authentication protocol comes to the aid to provide user authentication. The unique hardware and context that comes with the AR headsets is leveraged by the LGTM protocol with the intent of bringing human trust mechanisms into the digital world [11]. This is done to solve the problem of authentication, that too in a secured way. Users can authenticate in numerous conditions as LGTM works over wireless communication comprehensively. LGTM is designed at the heart of usability as users are required to perform only two actions: initiate and confirm. With the help of this protocol, users can share any content with each other face-to face just by selecting the person. It is assumed that each headset possesses a facial recognition model. LGTM protocol uses Diffie–Hellman key exchange in order to establish a secure key between two users. Confidentiality of the facial recognition parameters needs to be protected and thus symmetric key encryption is used by LGTM. For any technology, usability is the vital part. Apart from providing user satisfaction, it is also good for increasing security. It can lessen user errors that lead to issues related to security.

This protocol provides security against MiTM attacks. A combination of wireless localization and facial recognition is used by LGTM to reduce user-device authentication. With the help of LGTM, authenticating a wireless signal is made very simple. Through LGTM, the origin of wireless signal is merged with the location of the user's face. Implementation of LGTM can be done using various software, hardware, and technologies. The amount of security LGTM provides and its usefulness is expected to increase with time in the future of computing.

2.8.2 Doodle-based Authentication

The crux of the Doodle-based authentication lies in the amalgamation of AR with doodle passwords. In AR space, for the purpose of creation of doodle passwords, touch-gesture-recognition is used on smartphones. It is also studied that passwords based on doodles are very complex and thus difficult. Thus, doodle passwords provide security to the system. For this fusion of AR with doodle passwords, AR Drawing is chosen. AR Drawing is the Google creative lab implemented through AR. In order to create a password through doodle strokes, the screen of the smartphone is touched and moved in 3D space. This is done to draw a doodle shape which will be set as a password. Five doodles need to be drawn in total, out of which first four is for the registration and fifth doodle is for authentication. After the doodles are drawn by the user, coordinates of the drawn doodle are grabbed. It is done with the help of vector math libraries. Dynamic array lists store the coordinates of the registered doodles. Later, during the authentication process, these array lists are used.

For the purpose of authentication of the user, it is asked to draw the same doodle again which was drawn at the time of registration of the user. Then, the matching of the drawn doodle takes place with all the five doodles which were drawn by the user at the time of registration [14]. The difference is checked among the coordinates and on the basis of that extraction of the similarity is done. Matching of size and coordinates of two doodles will only lead to successful authentication. It is very difficult to crack doodle passwords as a very large number of doodles are possible in a given space. Thus, it can be proved that it is a very powerful authentication scheme.

2.8.3 Nod-to-Auth

With the growing popularity of AR/VR gadgets, researchers are focusing more on their authentication approaches. External input devices have been used in most password-based authentication techniques, which can be inconvenient to carry and operate. Furthermore, unlike smartphones and wearables, AR/VR headsets are frequently shared among users, necessitating more frequent authentication than personal devices. All of these necessitate novel AR/VR authentication methods that users can complete quickly and with minimal input.

In Nod-to-Auth protocol, an unique AR/VR authentication mechanism allows users to unlock their headsets with simple nodding gestures. Because this technique is based on the IMU sensors used in most AR/VR devices, it can be simply and inexpensively expanded to a wide range of devices.

IMU sensors are used to obtain biometric features from the head-neck portion of the wearer [12]. Average neck length, changes in neck length and head orientation during head gestures, and characteristics of the head gesture trajectory are examples of these features, which capture subtle but significant patterns that differ between users. A typical machine learning model is applied to identify users, which, according to a 10-person test, achieves average accuracy of 97.1%, 97.7%, 98.5%, and 99.1% for groups of 5, 4, 3, and 2, respectively.

2.8.4 MoveAR

Nowadays, AR headsets are easily available to consumers. Because these devices lack traditional input interfaces, it is necessary to investigate new techniques for providing security primitives such as user authentication. Given the multiple inertial sensors used by headsets to place users in their environment, this protocol allows for a continuous biometric authentication system based on the various ways in which people move their heads and engage with their virtual worlds. A group of users wearing an AR headset provides the samples of spatial and behavioral patterns. A variety of novel models and machine learning pipelines are based on these data that learn AR users' unique signature as they interact with the virtual world and AR objects. Several machine learning algorithms like k-nearest neighbors, random forest, SVM, and bag of symbolic-Fourier-approximation symbols (BOSS) for two different input data sets and parameters are evaluated. A 92.675% balanced accuracy score and an EER of 11% combined with adaptive boost random forest classifier and AR-specific pre-processing algorithms is used on the current dataset. This protocol proves that the users of AR head-mounted displays have to be profiled and authenticated on the movements and gestures of their heads [13].

The main goal of this protocol is to develop a novel and highly accurate model that can authenticate a wearer of a mixed reality headset using data collected from on-device sensors. Time series-based models and feature-based models are the two main categories of this protocol. The feature-based models accept a feature vector representation of each time series in the set S, whereas the time series-based models accept raw time series of the sensor data in the set S. As a result, time series-based approaches do not require a distinct feature vector to be extracted. In both cases, binary classifiers are created for each unique user in the dataset.

2.9 Comparison of Various Authentication Techniques

Several authentication techniques for AR systems are depicted. The survey is summarized in Table 2.1. It is clear from the survey that ZeTA and Nod-to-Auth can result in better performance with improved security features. Moreover, analysis of several authentication schemes on the basis of certain parameters are mentioned below and summarized in Table 2.2.

i. Source authentication

This parameter is used to validate the source ID which is originating the message. It is necessary to validate each broadcasted message at the receiver end.

ii. Cryptography/algorithms used

Cryptographic schemes are either symmetric or asymmetric digital signatures. And the algorithms are the set of rules upon which the authentication schemes are built. In our comparison, protocols are given the specific cryptography method it is using or the machine learning algorithms it uses.

iii. Security against user impersonation

While authenticating two AR devices, it is possible that an adversary can behave as a legitimate user and make itself authenticate to the target device.

iv. Security against DoS

Like anything, it is possible to perform DoS attacks in these authentication algorithms too. In our comparison table, we give different schemes if they are vulnerable to DoS attack.

v. Security against MiTM

As in most of the cases, either two AR devices authenticate with each other or the user authenticates him/herself with the AR device; in both the cases, authentication is happening over WSN, which is prone to MiTM attack. Our comparison table consists of a column that contains if certain schemes are vulnerable to this attack or not.

vi. Time synchronization

It is required that the sender does not have to send respective keys when sending the message. So, this security check helps the receiver to verify that, and in the table, some protocols need to share the keys so it is applicable to them, but some are using machine learning–based schemes where it is sometimes not required to share the keys, so they are given 'Not Applicable'.

TABLE 2.1

Survey of Different Authentication Techniques Used in AR

Performance Evaluation	In this scheme, researchers have explored the human-based computation. This scheme is secured in many ways, thus preventing impersonation and brute force.	This scheme employs the authentication using Diffie–Hellman Exchange algorithm for establishing shared-symmetric key and then final authenticity verification by facial recognition.	Nod-to-auth scheme unlocks the device by leveraging the IMU sensors to match the head gestures, which is based on Random Forest Classifier algorithm.	In this scheme, authentication is done on the basis of behavioral biometric, by extracting features and using machine learning pipeline to verify through IMU data.	This scheme follows Randomized-Posttest-Only Research design. This works by creating password in 3D space, hence avoid the impersonation issue.	This scheme consists of three major steps: EEG authentication, Eye-Tracking authentication, and Multimodal Fusion.
Year	2016	2016	2021	2021	2020	2019
Author with Reference	Gutmann et al. [9]	Gaebel et al. [11]	Wang et al. [12]	Bhalla et al. [13]	Wazir et al. [14]	Krishna et al. [10]
Techniques	ZeTA	LGTM	Nod-to-Auth	MoveAR	Doodle-Based Auth	Multimodal Biometric

TABLE 2.2

Comparison of Existing Authentication Schemes in AR

Security against MiTM	Yes	Yes	NA	NA	No	No
Security against DoS	No	No	NA	No	No	No
Security against User Impersonation	Yes	Yes	No	No	NA	Yes
Cryptography/ Algorithms		Diffie–Hellman key exchange	Random Forest Classifier	KNN Classifier	ARCore	SVM, Random Forest Classifier
Time Synchronization	Yes	Yes	Yes	NA	NA	NA
Source Authentication	Yes	Yes	Yes	Yes	Yes	Yes
Techniques	ZeTA	LGTM	Nod-to-Auth	MoveAR	Doodle-Based Auth	Multimodal Biometric

Options in Security Columns:

Yes: The protocol is secured against that type of attack.

No: The protocol is not secured against that type of attack.

NA: The type of attack is not applicable in that protocol because it may cause repetition of human body parts that is not possible in large sense and/or protocol is using the technique that does not feature any endpoint that can be used for that of attack.

2.10 Conclusion

In this chapter, the significance of authentication in AR systems has been elucidated. Furthermore, it is highly speculated to have numerous possible directions in the future. Various protocols have been discussed. These authentication protocols provide a clear idea and distinguish between authorized and unauthorized access. Comparison is very useful in determination and quantification of different protocols to check which one is better over the other. Each protocol has its own set of merits and demerits. The scope of the mentioned protocols is not confined till the written part. It further has the possibility of more research. The most challenging part in this domain is to construct a middleware which extensively supports the AR systems.

References

1. Gramfort, Alexandre, Luessi, Martin, Larson, Eric, Engemann, Denis A., Strohmeier, Daniel, Brodbeck, Christian, Parkkonen, Lauri, & Hämäläinen, Matti S.. (February 2014). MNE software for processing MEG and EEG data. *NeuroImage* 86, 446–460. doi:10.1016/j.neuroimage.2013.10.027

2. Saidin, Nor, Abd Halim, Noor, & Yahaya, Noraffandy. (2015). A review of research on augmented reality in education: Advantages and applications. *International Education Studies* 8. doi:10.5539/ies.v8n13p1

3. Wazir, W., Khattak, H. A., Almogren, A., Khan, M. A., & Din, I. U. (2020). Doodle-based authentication technique using augmented reality. *IEEE Access* 8. doi:10.1109/access.2019.2963543

4. Yu, K.Y., Yuen, T.H., Chow, S.S.M., Yiu, S.M., & Hui, L.C.K. (2012) PE(AR)2: Privacy-Enhanced Anonymous Authentication with Reputation and Revocation. In: Foresti, S., Yung, M., & Martinelli, F. (eds) *Computer Security – ESORICS 2012*. ESORICS 2012. Lecture Notes in Computer Science, vol 7459. Springer, Berlin, Heidelberg. doi:10.1007/978-3-642-33167-1_39

5. Gutmann, A., Renaud, K., Maguire, J., Mayer, P., Volkamer, M., Matsuura, K., & Muller-Quade, J. (2016). ZeTA-Zero-Trust Authentication: Relying on Innate Human Ability, Not Technology. In *2016 IEEE European Symposium on Security and Privacy (EuroS&P)*. doi:10.1109/eurosp.2016.35

6. Gaebel, E., Zhang, N., Lou, W., & Hou, Y. T. (2016). Looks Good To Me. In *Proceedings of the 6th International Workshop on Trustworthy Embedded Devices–TrustED '16*. doi:10.1145/2995289.2995295

7. Duezguen, Reyhan, Mayer, Peter, Das, Sanchari, & Volkamer, Melanie. (2020). Towards Secure and Usable Authentication for Augmented and Virtual Reality Head-Mounted Displays. *arXiv preprint arXiv:2007.11663*.

8. Krishna, V., Ding, Y., Xu, A., & Höllerer, T. (2019). Multimodal Biometric Authentication for VR/AR using EEG and Eye Tracking. In *Adjunct of the 2019 International Conference on Multimodal Interaction on – ICMI '19*. doi:10.1145/3351529.3360655

9. Wang, Xue, & Zhang, Yang. (2021). Nod to Auth: Fluent AR/VR Authentication with User Head-Neck Modeling. In *Extended Abstracts of the 2021 CHI Conference on Human Factors in Computing Systems* (pp. 1–7). doi:10.1145/3411763.3451769

10. Silva, Rodrigo, Oliveira, Jauvane, & Giraldi, G. (2003). Introduction to augmented reality. *National Laboratory for Scientific Computation*, 11, 1–11.

11. Carmigniani, Julie, & Furht, Borko. (2011). Augmented Reality: An Overview. In: Furht, Borko (ed) *Handbook of Augmented Reality*. doi:10.1007/978-1-4614-0064-6_1.

12. Carmigniani, Julie, Furht, Borko, Anisetti, Marco, Ceravolo, Paolo, Damiani, Ernesto, & Ivkovic, Misa. (2010). Augmented reality technologies, systems and applications. *Multimedia Tools and Applications* 51, 341–377. doi:10.1007/s11042-010-0660-6

13. Gandolfi, Enrico. (2018). Virtual Reality and Augmented Reality. In: Miedany, Yasser El (ed.) *Rheumatology Teaching* (pp. 403–427). Springer, Cham.

14. Cibilić, Iva, Posloncec-Petric, Vesna, & Tominić, Kristina. (2021). Implementing augmented reality in tourism. *Proceedings of the ICA* 4, 1–5. doi:10.5194/ica-proc-4-21-2021

15. Hull, Jonathan, Erol, Berna, Graham, Jamey, Ke, Qifa, Kishi, Hidenobu, Moraleda, Jorge, & Olst, Daniel. (2007). Paper-Based Augmented Reality. In *17th International Conference on Artificial Reality and Telexistence (ICAT 2007)* (pp. 205–209). doi:10.1109/ICAT.2007.49.

16. Sallow, Amira, & Younis, Mohamed. (2019). Augmented reality: A review. *Academic Journal of Nawroz University* 8, 76. doi:10.25007/ajnu.v8n3a399

17. Bhalla, Arman, Sluganovic, Ivo, Krawiecka, Klaudia, & Martinovic, Ivan. (2021). MoveAR: Continuous Biometric Authentication for Augmented Reality Headsets. In *Proceedings of the 7th ACM on Cyber-Physical System Security Workshop (CPSS '21)*. *Association for Computing Machinery* (pp. 41–52). New York, NY. doi:10.1145/3457339.3457983

18. Yuen, Steve Chi-Yin, Yaoyuneyong, Gallayanee, & Johnson, Erik (2011). Augmented reality: An overview and five directions for AR in education. *Journal of Educational Technology Development and Exchange (JETDE)* 4(1), 11. doi:10.18785/jetde.0401.10

19. Li, Xiao, Yi, Wen, Chi, Hung-Lin, Wang, Xiangyu, Chan, Albert P.C. (2018). A critical review of virtual and augmented reality (VR/AR) applications in construction safety. *Automation in Construction* 86, 150–162, ISSN 0926-5805. doi:10.1016/j.autcon.2017.11.003

20. Michalos, George, Karagiannis, Panagiotis, Makris, Sotiris, Tokçalar, Önder, & Chryssolouris, George. (2016). Augmented Reality (AR) applications for supporting human-robot interactive cooperation. *Procedia CIRP* 41, 370–375, ISSN 2212-8271. doi:10.1016/j.procir.2015.12.005

21. Sahu, Chandan K., Young, Crystal, & Rai, Rahul (2021). Artificial Intelligence (AI) in Augmented Reality (AR)-assisted manufacturing applications: A review. *International Journal of Production Research* 59(16), 4903–4959. doi:10.1080/00207543.2020.1859636

22. Livingston, M.A., Ai, Z., & Decker, J.W. (2019). Human Factors for Military Applications of Head-Worn Augmented Reality Displays. In Cassenti, D. (eds.) *Advances in Human Factors in Simulation and Modeling. AHFE 2018. Advances in Intelligent Systems and Computing* (vol. 780). Springer, Cham. doi:10.1007/978-3-319-94223-0_6

23. Hsieh, M. C., & Lee, J. J. (2018). Preliminary study of VR and AR applications in medical and healthcare education. *Journal of Nursing and Health Studies* 3(1). doi:10.21767/2574-2825.100030

24. Furtos, G., Tomoaia-Cotisel, M., Baldea, B., & Prejmerean, C. (2013). Development and characterization of new AR glass fiber-reinforced cements with potential medical applications. *Journal of Applied Polymer Science* 128(2), 1266–1273.

25. Han, D. I., & Jung, T. (2018). Identifying Tourist Requirements for Mobile AR Tourism Applications in Urban Heritage Tourism. In *Augmented Reality and Virtual Reality* (pp. 3–20). Springer, Cham. Chicago.

26. Han, D. I., Tom Dieck, M. C., & Jung, T. (2018). User experience model for augmented reality applications in urban heritage tourism. *Journal of Heritage Tourism* 13(1), 46–61.

27. Wheeler, G., Deng, S., Toussaint, N., Pushparajah, K., Schnabel, J. A., Simpson, J. M., & Gomez, A. (2018). Virtual interaction and visualisation of 3D medical imaging data with VTK and Unity. *Healthcare Technology Letters* 5(5), 148–153.

28. Collett, T. H. J., & MacDonald, B. A. (2006, March). Developer Oriented Visualisation of a Robot Program. In *Proceedings of the 1st ACM SIGCHI/SIGART Conference on Human-Robot Interaction* (pp. 49–56).

29. Parekh, P., Patel, S., Patel, N., & Shah, M. (2020). Systematic review and meta-analysis of augmented reality in medicine, retail, and games. *Visual Computing for Industry, Biomedicine, and Art* 3(1), 1–20.

30. Javornik, A. (2016). 'It's an Illusion, but It Looks Real!' consumer affective, cognitive and behavioural Responses to augmented reality applications. *Journal of Marketing Management* 32(9–10), 987–1011.

31. Wedel, M., Bigné, E., & Zhang, J. (2020). Virtual and augmented reality: Advancing research in consumer marketing. *International Journal of Research in Marketing* 37(3), 443–465.

32. Arth, C., Grasset, R., Gruber, L., Langlotz, T., Mulloni, A., & Wagner, D. (2015). The History of Mobile Augmented Reality. *arXiv preprint arXiv:1505.01319.*

33. Berryman, D. R. (2012). Augmented reality: A review. *Medical Reference Services Quarterly* 31(2), 212–218.

34. Omidshafiei, Shayegan, Agha-Mohammadi, Ali-Akbar, Fan Chen, Yu, Kemal Ure, Nazim et al. (December 2016). Measurable augmented reality for prototyping cyberphysical systems: A robotics platform to aid the hardware prototyping and performance testing of algorithms. *IEEE Control Systems* 36(6), 65–87. © 2016 Institute of Electrical and Electronics Engineers (IEEE)

35. Lukman Khalid, C. M., Fathi, M. S., & Mohamed, Z. (2014). Integration of Cyber-Physical Systems Technology with Augmented Reality in the Pre-Construction Stage. In *2014 2nd International Conference on Technology, Informatics, Management, Engineering & Environment* (pp. 151–156). doi:10.1109/TIME-E.2014.7011609

3

An Automated Cover Text Selection System for Text Steganography Algorithms

Anandaprova Majumder

Dr. B. C. Roy Engineering College, Durgapur, India

Suvamoy Changder

National Institute of Technology, Durgapur, India

CONTENTS

DOI: 10.1201/9781003241348-3

3.1 Introduction

Steganography is about hiding some secret message within any cover media file, without any doubt on its very existence by any unwanted recipient. Cryptography just like the soul sister of steganography follows a different way of data encryption and makes a message indecipherable by any unwanted receiver but cannot hide the existence of the ongoing secret communication. Steganography and cryptography are treated as two strong pillars of secret communications instead of their solely distinct and different nature. With modernization of society, need for digitization is increasing in our everyday life and the requirement of data securing and secret communication has become very essential [1, 2]. As an after effect, it has attracted the attention of researchers as well. The charm of steganography lies in the undetectable presence of the secret message inside the cover media. The chosen cover file is either modified or mapped or synthesized with an aim of generation of an innocuous stego-cover [3] that is to be shared through any public communication channel. The communication channel is considered to be a lossless one and the message is received as it is sent from the sender end. The receiver upon receiving the stego-object, applies the decryption algorithm to regenerate the hidden secret message from it [4]. The idea of steganography lies in establishing a secure and safe data transmission in an unidentifiable manner and to stay out of attracting any suspicious attention.

So, it is about keeping the outsiders from even knowing the existence of information, not only to transform the secret data in a special form that is not understandable. If the presence of secret data is suspected in a cover media file, the steganography method fails. The first documented evidence of steganography is the historic story about the shaved head slaves of Herodotus.

In modern steganography algorithms, the cover file is updated in any of the three ways like cover modification, cover selection, and cover synthesis [3, 5]. If cover modification is followed, encryption algorithm modifies the bits of cover file in an unnoticeable fashion. If cover selection is used, the cover file is no way modified; some kind of mapping technique is used for data hiding. If cover synthesis is used for data hiding, a new cover file is generated as output of encryption algorithm. A specific steganography technique can be named as text steganography, if the chosen cover file is of type text. Text steganography is considered as the toughest type of steganography because of lack of repetitive or redundant information present in a text file and it is hard to hide the modifications in original text files [6]. In the General Text Steganography model, as shown in Figure 3.1, embedding function works at the sender end and takes the secret text, cover text, and some optional key as

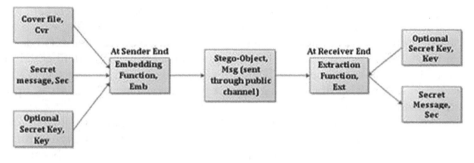

FIGURE 3.1
General steganography model.

input to generate the stego-cover and it is sent through the communication channel. The receiver applies the extraction algorithm on the received stego-cover and regenerates the hidden secret message sent by the sender. Numerous amount of steganography methods have been applied on different types of cover media files such as text [7, 8], image [9–12], audio [13], and video files [1, 14].

This chapter presents a novel idea for automated selection of most accurate cover text from the available database of large number of cover text files, which can be any text paragraph of sufficient length available in newspaper archives or online blogs, etc. and can be used for hiding the chosen secret text in it. The proposed method is to be applied at the sender side data encryption end, for finding the most applicable cover text file from thousands of digitally available resources, which will hide the chosen secret message innocuously without generating any suspicion. The process of proper cover text selection needs to be automated for faster application compared with existing manual process. With availability of different text steganography algorithms, if an accurate cover file can be selected fast that will make the encryption system well accepted.

For proper explanation of this research, we have elaborately discussed different parts of it in different sections. In the next section, we have continued the introductory discussions. Related literature study regarding existing text steganography methods and natural language processing using machine learning is discussed in the following Section 3.2. From that study, we have derived our motivation and moved further with our contribution of proposed automatic cover text detection system in Section 3.3. Experimental results and some comparative analysis have been discussed further in Section 3.4. In Section 3.5, we have finally concluded our chapter and cited the related references in the following Reference section.

3.2 Related Literature Study

Steganography is segmented under different classes depending on types of carrier media files used like image, text, audio, video files etc. However, how much redundancy is there in the media files or how much similar the cover file looks like after embedding are the key issues that measure the effectiveness of the stego-algorithm. Lesser the changes made on the original cover file, more will be the effectiveness of the applied stego-algo. Different ways of embedding has come up by following different embedding techniques in steganography, like insertion-based techniques, substitution-based, and generation-based techniques. Insertion-based methods selects the location in the cover file, which are the ignorable areas and where to hide the message. Substitution-based methods use redundant data inside the cover media files for hiding secret messages and deliberately interchange some bits of the cover file with the bits of secret message for hiding. Generation- or synthesis-based methods synthesize a new cover of its own instead of anyway updating any existing cover media file.

Text steganography method incorporates different processes like format changing of a pre-existing text file, changing different words within a pre-existing text, generating readable texts following context-free grammar or some random sequence, etc. Because of the presence of enough redundant information, secret data are easier to be hidden in image, audio, or video file, whereas it is not that easy in case of text steganography. The text document structure is same as that is observable while the document structure is different from the viewable one in case of an image. From the storage point of view, text file will need much small memory space and it is much faster to communicate compared with any other types of steganographic methods using some other kind of media files. Overall, the method of text steganography is classified into three types: Format or structure-based methods, random and statistical generation-based methods, and linguistic methods. Now, we will be studying different well-established text steganography methods that have been proposed in recent days.

3.2.1 Format-Based Methods

Physical format [15, 16] of text is altered to conceal the secret information in case of application of format-based methods. Certain shortcomings are identified in this method. Opening of stego file with a word processor will easily identify the spelling mistakes and extra blank spaces. Font size changing may also raise suspicion to a person reading the document. Updated parts of the text would become quite visible if a comparison is done between the original plain text and suspected steganographic text. The format-based methods are well established because of the ease of application.

3.2.2 Random and Statistical Generation-Based Methods

Generating own cover texts is often found out to be a better practice followed by practitioners of steganography in order to avoid comparison with existing plain texts. One followed method can be hiding of information with the help of random character sequences. The statistical properties [17, 18] of length of the word and letter frequencies are used for example for creating words with an appearance of having same statistical properties as actual words in the followed language in a different method. Synthesizing an innocent-looking cover text file is not an easier approach in comparison with the format-based methods.

3.2.3 Linguistic Steganography

Different linguistic properties of created and updated text are specifically considered in case of linguistic steganography. In many cases, linguistic structure like the blank space is used for secret data hiding. Context-free grammar [19] creates tree-like structures. Those structures are used for hiding the secret bits where left branch may be used for hiding '0' and right branch for hiding '1'. A grammar in GNF can be used for the purpose too, where to hide bit 0, the first choice in a production is used and the second choice is used to hide the bit 1. Certain shortcomings are there in this case too. First, a lot of repetition in text representation can be caused by a small grammar. Second, there may be lack in semantic structure, although the text is syntactically flawless, that is the meaning may get altered and this may result in a string of out-of-context sentences.

3.2.4 Existing Text Steganography Methods

In the following segment, some of the well-known text concealing techniques, that is the approaches of text steganography, have been discussed.

3.2.4.1 Line Shift

Vertical shifting of the text lines to some degree is used for data hiding in line shift method of text steganography. Two unmarked control lines are used with a marked line one on two sides of it for finding the orientation of movement of the marked line. For hiding bit 0 and 1, a line is moved up and down accordingly. The shifting of a line toward upward or downward direction is determined by calculating the distance of the center of marked line and its control lines [7]. With retyping of a text or with the use of an optical character recognition program, the concealed message would get changed, hence meaningless. The distances are also measurable with the help of special apparatus.

3.2.4.2 Word Shift

Horizontal shifting of the words can be done to either of the directions, right or left to hide bit 0 or 1 accordingly, for secret data hiding in this method. Correlation methods are used to detect the word shifts. It treats a profile as a waveform and checks whether it is generated from a waveform whose middle block is moved toward left or right [6, 8]. This method is possibly less identified as it is quite a common practice to change the distance between words to fill a line. But for a person with the knowledge of the algorithm of distances, it is quite easier to compare the applied algorithm along with the stego-text and find out if there is something concealed in it.

3.2.4.3 Syntactic Method

To hide data bits 0 and 1, different punctuation marks are used in this method such as full stop (.), comma (,), and semicolon (;) [7]. Proper identification of correct locations to insert different punctuation marks is required in this method. Therefore, it needs a much careful approach to keep the readers from identifying wrong use of punctuations symbols.

3.2.4.4 White Steg

This method conceals secret data using white spaces. Three methods of data hiding are found by use of white spaces. Single space is placed for hiding bit 0 and two spaces for hiding bit 1 at the end of each terminal character in case of inter-sentence spacing. A predefined number of blank spaces are inserted at the end of every line in case of end of line spaces [6]. One space after a word is used for representing bit 0, and to represent bit 1, two spaces are used after a word in case of inter-word spacing technique. But, inconsistent use of white space may attract attention of outsider.

3.2.4.5 Feature Coding

One or more features or properties of the text can be changed or altered to enable the sender to hide secret data using this method. A document is checked by a parser and all the features are picked out that it can be used for information hiding [7]. For example, displacement in the point in letters '*i*' and '*j*' can be made, change in the length of horizontal strike mark in letters '*f*' and '*t*' can take place, or height of letters '*b*', '*d*', '*h*', etc. can be shortened or elongated. There is a shortcoming of this method too. If a retyping is done or an optical character recognition system is used, the concealed data would get easily corrupted and destroyed accordingly.

3.2.4.6 Word Mapping

A secret data is encrypted by following the word mapping technique with the use of genetic operator crossover. The cipher text obtained as a result is then embedded considering a couple of bits at a time in a cover file by pushing some blank spaces between words of odd or even length using a specified mapping technique [8]. In another file, the locations of embedding are saved and finally shared with the receiver along with the stego file.

3.2.4.7 Microsoft Word Document

Text segments in a document are degenerated in this method, imitating to be the manuscript of a writer with poor writing capability. Secret message is embedded according to the choice of degenerations which are then revised later with changes being tracked [6, 7]. Data hiding is represented in such a way that stego document appears to be the output of a collaborative writing.

3.2.4.8 Text Steganography Using AITSTEG

A recent innovative hidden data transmission technique for sending text messages in social media has been proposed in 2018 by Jing Zhang et al. in the field of text steganography. It provides end-to-end security during data transmission between the end users by using social media or sms. This method shows a comparative better result in terms of robustness, better embedding capacity, nature of being undetectable, and security with respect to the previously existing algorithms as shown in [20].

3.2.4.9 Coverless Plain Text Steganography Based on Character Features

Coverless text steganography is a research hot topic in recent days because zero modification on the cover increases the concealment of a system. A coverless plain text steganography system based on character features was proposed in 2019 by Kaixi Wang et al. as shown in [21]. The parity of Chinese characters stroke numbers and its statistical characteristics has been analyzed by it. The researchers have employed parity to express a binary digit and its combination to express binary bit string.

3.2.4.10 Steganography via e-Books by Rearranging CSS Files

In 2019, Da Chun Wu et al. has proposed a novel data-hiding method into the e-book with EPUB format for steganographic applications. Secret message bits are hidden into the cascading style sheets without changing the appearance of the e-book. The lexicographical orders of the selectors and their declarations in the cascaded style sheet rules are used to hide the message bits into the cascading style sheets as referred in [22].

Different novel steganography algorithms have been proposed in recent years and have been referred in [23–28]. In this novel proposed work, we have used natural language processing for analyzing sample cover text files and depending on the requirement of the chosen secret text, an accurate cover text is selected for data hiding. In this process of automated cover text selection, fuzzy clustering algorithm is used.

3.2.5 Motivation Derived from Literature Survey

Studying different contributions in the related research domain, we have gathered knowledge about different types of well-existing text steganography methods. We have also got the idea of the properties that must belong to the cover texts able to hide the given secret texts in an unidentifiable manner. The trial-and-error process of manual cover text selection is time consuming and the same process can be made faster by automation. A cover text database can be made out of huge set of digital text resources available online and some natural language processing technique can be used for selection of the most suitable one for hiding the respective secret text.

3.3 Proposed Automated Cover Text Selection System

With an emergent need of faster system output, the process of selection of cover text needs to be automated. It is a manual process even in recent days and hence is time consuming. The proposed system is an approach toward making the process of cover text selection, an automated one. The method will select the most accurate cover text to hide a given secret text using the given text steganography algorithm. The proposed model is a well-diffused combination of six component sub-modules that establish automated cover selection at sender end and the sub-modules are shown in Figure 3.2 as follows.

All these six sub-modules are working with different functions in the background, and by the application of these functions, the proposed automated system is generated which is applied at the sender end. In the sender end, the proposed system takes the secret message and the stego-algorithm to be implemented as input. After analyzing the secret text, the system generates the threshold secret vector, which is followed by the requirement analysis of the chosen stego-algorithm. With the help of natural language processing, cover text files are analyzed with an aim to produce the cover vector. The fuzzy clustering algorithm is applied next to select the most accurate cover text out of the database that conceals the input secret text innocuously. The proposed method can be well explained with the help of the flowchart as shown in Figure 3.3.

3.3.1 Cover Text Data Set Preparation

In application of text steganography, cover text is the media that conceals the secret data in such a way, so that the presence of secret text remains unnoticed to unintended recipients. Normally, any sufficiently long cover text can be chosen as the hiding text media.

For our proposed automated system, for application of the same, a set of such 50 cover texts has been collected, which are nothing but innocent-looking, publicly available, large

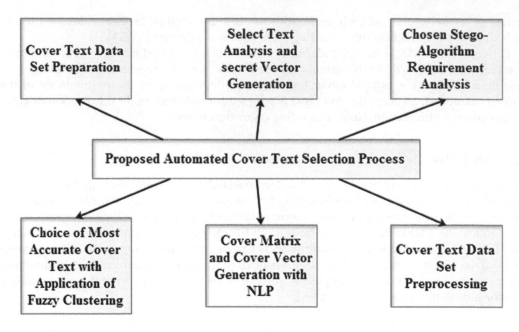

FIGURE 3.2
The component sub-modules of proposed automated cover selection process.

enough English paragraphs, obtained from online blogs, newspaper articles, story book chapters, etc. A snapshot of the cover data set is illustrated in Figure 3.3.

3.3.2 Secret Text Analysis and Secret Vector Generation

Secret text that is to be shared by the sender is hidden inside the cover text and in order to do that, analysis of secret text is first required for automated selection of the accurate cover text. After the secret message is entered as input in the sender encryption system, the process of secret vector generation starts following these below explained steps.

- Secret text, Sec_Text, is converted from alphanumeric form to an equivalent ASCII form and followed to that, the ASCII pattern is converted to binary bit pattern, which is the raw data to be further analyzed by the system.
- Secret vector is a row vector that is generated by calculating eight different elemental values from the binary bit pattern of the secret text as follows:

3.3.2.1 Capacity

This attribute counts the total number of bits or the length of the bit pattern obtained from the secret text after converting it to binary. The first component of the secret vector Sec_Vect is capacity (V_1), which is calculated from the following formula.

$$V_1 = \sum_{i=0}^{k} n_i \qquad (3.1)$$

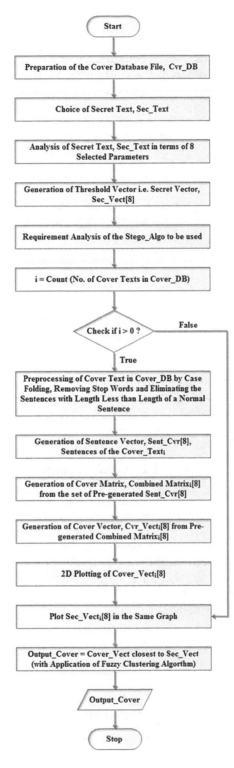

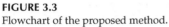

FIGURE 3.3
Flowchart of the proposed method.

3.3.2.2 Frequency Count

This attribute is a Boolean value and it is stored as a second component of the secret vector Sec_Vect and denoted as V_2. This field V_2 contains a value true, that is 1, if count of 0s and 1s are equal in the bit stream; otherwise, it contains a Boolean value false, that is 0. This gives the value for V_2, next component of secret vector Sec_Vect, and is shown in following formula.

$$V_2 = \text{Either True or False} \tag{3.2}$$

3.3.2.3 Percentage Score of '0'

If total count of '0's in the secret bit stream is known, it will help in finding an accurate cover text that will be able to hide that many '0's. This attribute gives the value of percentage score of '0' over total count of total number of bits in the bit stream and denoted as V_3, 3rd component of secret vector Sec_Vect and is shown by the following formula.

$$V_3 = \frac{\text{Total count of '0's}}{\text{Total no. of bits in the bit stream}} \tag{3.3}$$

3.3.2.4 Hamming Weight

The value of hamming weight will give the number of 1's in the secret bit stream, and this information is required for selection of a capable cover text. This attribute gives the value of total number of '1's in the total bit stream and denoted as V_4, the 4th component of secret vector Sec_Vect.

$$V_4 = \sum \text{Total count of '1's in binary form of sec\_Text} \tag{3.4}$$

3.3.2.5 Chi-Square Test of Independence for Presence of '0'

This attribute calculates the relative presence of '0' in the secret binary bit stream. The total length of stream is obtained from attribute V_1 and half of the value is considered as the expected count for presence of '0'. Scanning the bit stream once again gives the observed count for presence of '0'. Putting them in the following formula, the Chi-square value of the same is generated to fill the element value V_5.

$$V_5 = \chi^2_{[0]} = \sum \frac{(\text{observed} - \text{expected})^2}{\text{expected}} \tag{3.5}$$

3.3.2.6 Chi-Square Test of Independence for Presence of '1'

This attribute calculates the relative presence of '1' in the secret binary bit stream. The total length of stream is obtained from attribute V_1 and half of the value is considered as the expected count for presence of '1'. Scanning the bit stream once again gives the observed count for presence of '1'. Putting them in the following formula, the Chi-square value of the same is generated, to fill the element value V_6.

$$V_6 = \chi^2_{[1]} = \sum \frac{(\text{observed} - \text{expected})^2}{\text{expected}} \qquad (3.6)$$

3.3.2.7 Probability of Occurrence of '0' after '1'

This attribute V_7 measures the number of times '0' appears after '1' in the converted bit pattern originated from the secret text. This probability value gives the count of occurrence of pattern "10" in the bit pattern generated from Sec_Text. This attribute comes into picture in case of stego-algorithms that consider quadruple categorization, hence hides two bits at a time and those two bits can be arranged in four different ways.

3.3.2.8 Probability of Occurrence of '1' after '0'

This attribute, V_8, measures the number of times '1' appears after '0' in the converted bit pattern originated from the cover text. This probability value gives the count of occurrence of pattern "01" in the bit pattern generated from Sec_Text. This attribute comes into picture in case of stego-algorithms that consider quadruple categorization and hides two bits at a time, that is four different ways.

- Combining these eight components, the secret vector Sec_Vect is generated. Sec_Vect gives the threshold vector that is needed to be checked with the vectors obtained from different cover files of the cover text set.

Sec_Vect = V_1 V_2 V_3 V_4 V_5 V_6 V_7 V_8

3.3.3 Stego-Algorithm Requirement Analysis

Requirements of different text steganography algorithms to hide a given secret text can be analyzed for finding an accurate cover text for solving the purpose. The requirement of the stego-algorithms for hiding any secret text is to be analyzed beforehand, for getting a feedback of what kind of cover file will be best suited for application of the same. A set of six well-established and popular text steganography algorithms have been used for implementing the prototype model of the proposed work that generates an automated system of cover file selection. The requirements of all the chosen six algorithms have been studied and the cover data set has been analyzed accordingly with the help of natural language processing module of machine learning. The six most popular text steganography algorithms that we have considered for implementation purpose are:

1. Text steganography with synonym substitution
2. Text steganography using punctuation marks
3. Text steganography using feature coding
4. Text steganography using white space method
5. Text steganography based on curves in English language characters
6. Text steganography by data hiding in specific characters in different sentences

For all the six algorithms, their requirements for the kind of cover text they need to hide a secret message most accurately are studied. This study is important for the accuracy of the proposed system.

3.3.4 Cover Text Data Set Preprocessing

Before analysis of text with the help of machine learning modules, first a method of preprocessing is applied on the cover text files of the set in order to increase the accuracy in cover text finding. First, case folding and stop word elimination is done on the cover file. After that, normalized lengths of the sentences are calculated by dividing their lengths with the length of the largest sentence available in the corresponding cover file. Sentences with the normalized length more than a prespecified threshold value are considered for data hiding. If α_i denotes the normalized length of the ith sentence, it is calculated by dividing n_i with n_1 where n_i is the number of words in the ith sentence and n_1 denotes the number of words in the longest sentence in the file.

$$\alpha_i = n_i / n_l \tag{3.7}$$

3.3.5 Cover Matrix and Cover Vector Generation with NLP

Cover files are analyzed using natural language processing, and according to the chosen algorithm, the tokens are chosen from the cover files that are to be treated to hide '0's and '1's accordingly. Those tokens are counted accordingly and different rows of sentence vector, Sent_Cvr [8], are generated from each sentences of the corresponding cover file. The number of rows in the cover matrix will be equal to the number of sentences in the cover file. The generated sentence vectors are arranged in row-wise manner in the sequence of their occurrence and that gives the generated cover matrix, Combined_Matrix [8]. Hence, the entry in location Combined_Matrix$_{[i,j]}$ will be the ith attribute value obtained from the jth sentence of the cover file. Summing up the entries in the cover matrix in column-wise manner, the cover vector, Cover_Vect [8], is generated.

Sent_Cvr [8] = $\quad\quad S_1 \quad S_2 \quad S_3 \quad S_4 \quad S_5 \quad S_6 \quad S_7 \quad S_8$

Combined_Matrix [8] = $\quad M_{[1]} = S_{1.1} \quad M_{[1,2]} = S_{2.1} \quad M_{[1,3]} = S_{3.1} \quad \cdots\cdots$

$\cdots\cdots$

$\quad\quad M_{[i,1]} = S_{1.i} \quad M_{[i,2]} = S_{2.i} \quad \cdots\cdots \quad\quad\quad M_{[i,i]} = S_{j.i}$

Cover_Vect [8] = $\quad\quad C_1 = M_{[1]} + M_{[1,2]} + M_{[1,3]} + \cdots + M_{[i,1]} \quad C_2 \ C_3 \ C_4 \ C_5 \ C_6 \ C_7 \ C_8$

3.3.6 Choice of the Most Accurate Cover Text with Application of Fuzzy Clustering

Last step of the proposed system is the choice of most accurate cover text file from the cover text data set and the corresponding cover file is obtained as output of the proposed system. For implementation of this automated system, analysis of the cover files as per the requirement of the stego-algorithm is carried out and a cover vector is also generated in the previous step. The cover vector is matched with pregenerated secret vector, that is the threshold vector. The cover vector that is most accurate for data hiding is chosen from the cover file set. The accuracy is identified by analyzing its difference from the threshold vector value, which is the minimum requirement to be fulfilled for hiding the given secret text following the given stego-algorithm most innocuously. The distance between the threshold vector and the cover vectors is measured, and they are plotted in a two-dimensional plot to check out the closest cover text that can hide the secret text in the current scenario with application of fuzzy clustering algorithm. Ultimately, the cover file that is in the closest regime is chosen as the output of the automated cover text selection system.

3.4 Experimental Results and Analysis

3.4.1 Experimental Results

The proposed automated cover text selection system has been implemented in this piece of research, in which a cover text data set has been made consisting of 50 different text data samples. This sample cover data set has been used as training data in backend, using which the experiment has been carried out. The snapshot of the sample cover data set file is shown in Figure 3.4. Experiments have been carried out in this case by selecting different secret text and different stego-algorithms at different times and the system works well in every case.

Six different text steganography algorithms have been considered in this proposed experimental work as the sample stego-algorithm set for implementation and for an easy user interactive system at the sender encryption end. As per the proposed implemented system, the front end gives six options of different existing algorithms, out of which one is to be selected at a time. The system asks for the secret message to be entered and any alphanumeric sample can be considered as secret text. After entering the secret data, it gets analyzed with the help of natural language processing and the corresponding secret vector is generated, which will work as threshold vector. After that, the chosen algorithm is checked by the system and requirement analysis for the same is done. Then, the system analyzes cover text files one by one from the data set and generates their corresponding

```
1] Weeks after reports of a mysterious new virus began to emerge in the
central Chinese province of Hubei, the authorities there suddenly changed
how they determined who was infected.It led to a significant spike in
numbers - only because doctors are now counting patients who are diagnosed
in a clinic and not just those who have taken the test.But in those early
days, the rapid spread of the virus through the city of Wuhan, combined
with a shortage of hospital beds, meant some didn't get the treatment they
needed. Two Wuhan residents told the BBC about the harrowing experience of
trying to get care for their loved-ones in a city overwhelmed by sickness.A
few Integrated gasification combined cycle (IGCC) coal-fired power plants
have been built with coal gasification. Although they burn coal more
efficiently and therefore emit less pollution, the technology has not
generally proved economically viable for coal, except possibly in Japan
although this is controversial. [145] [146] Carbon capture and storage
Although still being intensively researched and considered economically
viable for some uses other than with coal; carbon capture and storage has
been tested at the Petra Nova and Boundary Dam coal-fired power plants and
has been found to be technically feasible but not economically viable for
use with coal, due to reductions in the cost of solar PV technology.[147]
```

FIGURE 3.4
Sample cover database file, Cover_DB.

cover vectors as per the requirement of stego-algorithm to be implemented. The cover vectors along with the threshold vector are plotted next and fuzzy clustering algorithm is used for finding the nearest match. The cover file with cover vector of nearest match with the threshold vector is considered as the most accurate cover file and considered for data hiding using the corresponding stego-algorithm. Initially, all the cover vector points are considered under the general cover point cluster. Then, Euclidian distance of secret vector point in the graph is calculated from all the cover vector points, and depending on the distance, the farthest points are eliminated from the cluster of possible cover points. The cluster of possible cover points represent the cover text files that are able to hide the chosen secret text. Ultimately, the cover file that is the nearest match with respect to the chosen cover text is the output of the automated selection system.

All the algorithms for different secret text samples are tested and the automatic cover file detection system works well in all the cases. Some sample example output files are shown here. The system is tested for the algorithm of text steganography with punctuation and the secret text chosen for hiding is "Hello". The secret text first goes through the text preprocessing step. Then, the eight components of secret vector are calculated and accordingly the secret vector is generated. Now, the requirements for the text steganography algorithm with punctuation are studied and some punctuations are used for hiding '0' and some others are used for hiding '1'. Accordingly, the cover vectors are generated from the cover files using that information. Finally, the cover vectors are matched with secret vectors and the cover vector with nearest value to the secret vector is chosen as output and comes as output in the front end, which is shown in Figure 3.5 and 3.6.

Figures 3.7 and 3.8 show the interface for application of text steganography algorithm with feature coding. In this case, the special features of some alphabets are used for data hiding, and our system, after text analysis, considers the same for hiding '1' and '0'. The secret message used here for data hiding is "Hi there".

Another sample interface to test the implementation of text steganography algorithm using English alphabets using curves is shown in Figures 3.9 and 3.10. Letters with curves are used for hiding '0' and without curves are used for hiding '1'. Accordingly, the system generates the cover vector, and secret vector generated from the message "hi there" is matched with that to finally output the most accurate cover text.

The process used in this research can be used for automated selection of cover text file from any sample cover text dataset and can be implemented for application of any text steganography algorithms.

3.4.2 Comparative Analysis of Experimental Data

The proposed method is experimentally applied and it has been analyzed considering a limited size cover data set and a limited number of text steganography algorithms. The same process can be applied for any cover data set size and any number of text steganography algorithms. The automated process of cover data selection makes the hiding procedure much faster compared with existing manual methods. The comparison of the existing manual cover text selection process with the proposed automated cover text selection process can be represented by the following chart as shown in Table 3.1.

By analyzing the comparison data as shown in Table 3.1, it can be pretty well said that the proposed automated method shows a much better success rate in choosing the cover text for secret data hiding and the time needed in the process is also much less than the manual process of the same.

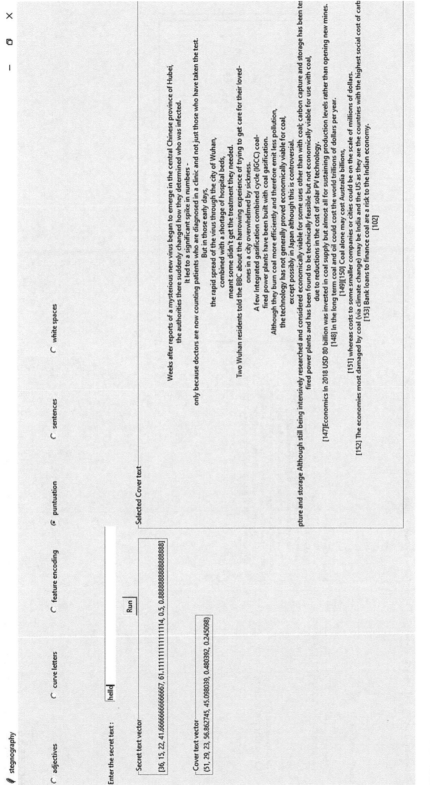

FIGURE 3.5
Application of text steganography with punctuations.

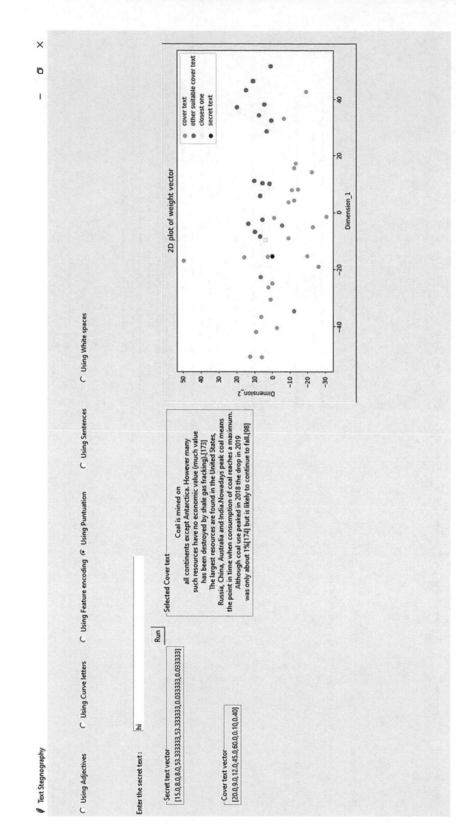

FIGURE 3.6
Interface for application of text steganography with punctuations with graph.

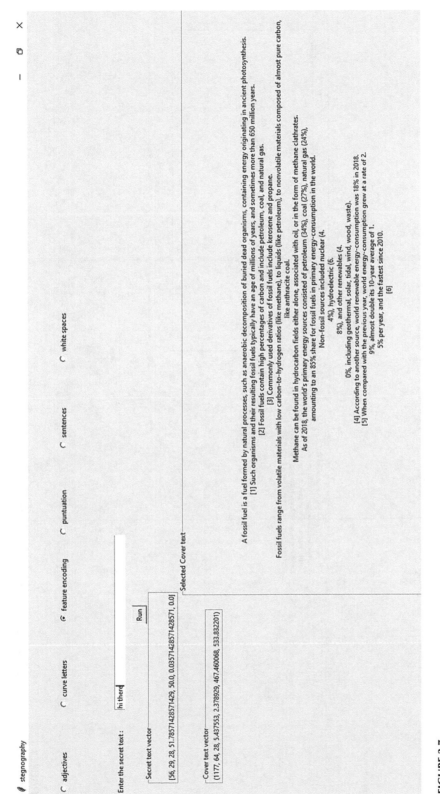

FIGURE 3.7
Interface for application of text steganography with feature coding.

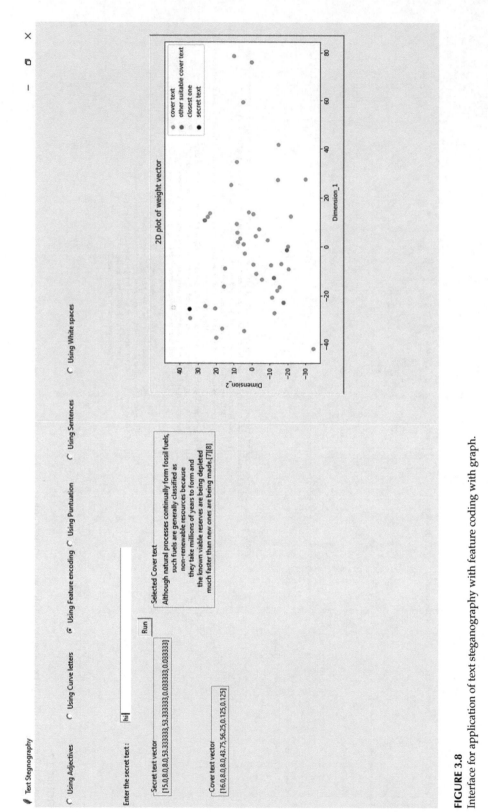

FIGURE 3.8
Interface for application of text steganography with feature coding with graph.

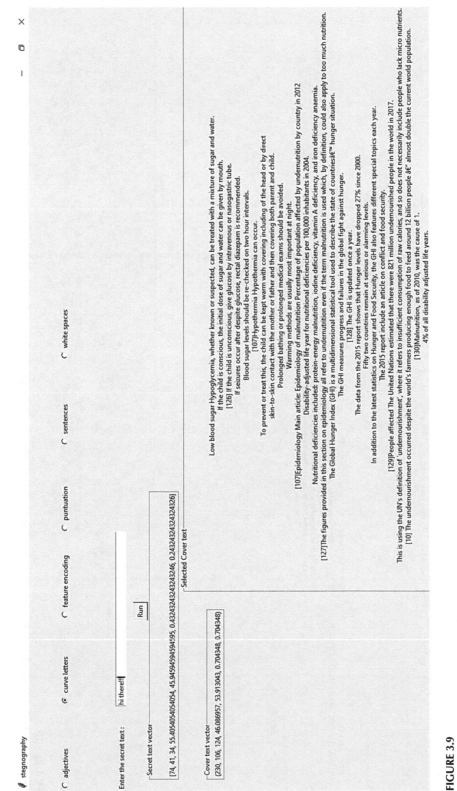

FIGURE 3.9

Interface for application of data hiding based on curves in English alphabet.

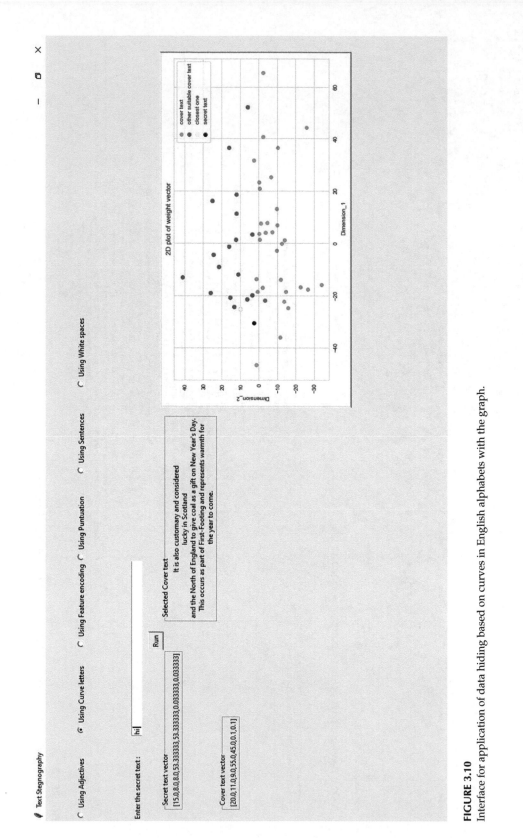

FIGURE 3.10
Interface for application of data hiding based on curves in English alphabets with the graph.

TABLE 3.1

Comparison of Manual and Automated Cover Text Selection Process

Text Steganography Algorithm Used	Manual Cover Text Selection Success Rate (%)	Proposed Automated Cover Text Selection Success Rate (%)	Time Required in Automated Process Compared with Manual Process (%)
Text steganography with synonym substitution	98.5	100	<50
Text steganography using punctuation marks	98.8	100	<60
Text steganography using feature coding	99	100	<50
Text steganography using white space method	99.2	100	<60
Text steganography by data hiding in specific characters in different sentences	99.5	100	<60
Text steganography based on curves in English language characters	99	100	<50

3.4.3 Time Complexity of the Proposed Method

The proposed method works in three basic steps and the time complexity of the proposed automated cover selection process can be found from the total time needed for completion of the same and it is the summation of the time needed by the three sub-steps of it. The total time (T_{total}) needed for completion of proposed method can be obtained by summing up the time ($T_{\text{pre-processing}}$) needed for preprocessing (cover database generation+ requirement analysis of algorithm to be applied), the time ($T_{\text{secret-vector generation}}$) needed for secret vector generation, the time ($T_{\text{cover-vector generation}}$) needed for cover vector generation, and the time (T_{matching}) needed for matching the secret vector with the cover vector to find the most suitable one. It can be represented by the following equation:

$$T_{\text{total}} = T_{\text{pre-processing}} + T_{\text{secret-vector generation}} + T_{\text{cover-vector generation}} + T_{\text{matching}}$$

The preprocessing time is fixed and it is applicable only once at the starting of the algorithm. Similarly, the secret vector, that is the threshold vector generation and the cover vector generation process needs a fixed time. Moreover, the time needed for matching the secret vector with the cover vector is also fixed. Hence, it can be said that total time needed for the proposed algorithm to work is fixed and the time complexity can be represented in terms of $O(1)$ as it is not variable in terms of the input size.

3.5 Conclusion and Future Research

An automated cover text selection system has been proposed in this chapter with an application of fuzzy clustering method. Analysis of cover text files has been done with natural language processing module of machine learning, and the requirement of different text steganography algorithms have also been studied. Finally, the generated cover vector is matched with the pregenerated secret vector and the cover files that produce cover vector with larger value compared with the threshold vector value are chosen. After plotting the cover text vectors and secret vector in the two-dimensional plot, distance of the cover vectors is measured with respect to the secret vector, and the cover vector having minimum distance from the secret vector is found using fuzzy clustering approach. The need for such a system arises from the fact that there are so many digital text data available publicly nowadays, from which any one can be chosen as a cover file, but the most accurate one is to be selected that will produce a more innocent-looking cover file. In addition, the selection process can be very fast while being automated compared with the existing manual process.

The proposed model has been tested with a cover file database of 50 sample cover texts in it and six well-existing text steganography algorithms have also been considered. Using fuzzy clustering algorithm depending on the secret text to be hidden and the stego-algorithm to be used for hiding, the most accurate cover text that can generate the most innocuous stego-text is obtained as the proposed system's autogenerated output.

The proposed novel method can be easily updated for selection of a digital cover text of any regional language as future research scope.

References

1. Patel, R., Lad, K., Patel, M. et al. A hybrid DST-SBPNRM approach for compressed video steganography. *Multimedia Systems*, 2021. https://doi.org/10.1007/s00530-020-00735-9
2. Zhu, Z., Ying, Q., Qian, Z. et al. Steganography in animated emoji using self reference. *Multimedia Systems*, 2021. https://doi.org/10.1007/s00530-020-00723-z
3. Liu, J., Ke, Y., Zhang, Z. et al., Recent advances of image steganography with generative adversarial networks. *IEEE Access*, 8, 60575–60597, 2020. https://doi.org/10.1109/ACCESS.2020.2983175
4. Uljarević, D., Veinović, M., Kunjadić, G. et al. A new way of covert communication by steganography via JPEG images within a Microsoft Word document. *Multimedia Systems*, 23, 333–341, 2017. https://doi.org/10.1007/s00530-015-0492-3
5. Yeung, Y., Lu, W., Xue, Y. et al. Secure binary image steganography based on LTP distortion minimization. *Multimedia Tools and Applications*, 78, 25079–25100, 2019. https://doi.org/10.1007/s11042-019-7731-0
6. Majumder, A., Changder, S. A novel approach for text steganography: Generating text summary using reflection symmetry. *Procedia Technology*, 10, 112–120, 2013.
7. Majumder, A., Changder, S., Debnath, N. C. A new text steganography method based on sudoku puzzle generation. *Lecture Notes in Electrical Engineering* (pp. 961–972), Sep. 2019.
8. Maji, G., Mandal, S. A forward email based high capacity text steganography technique using a randomized and indexed word dictionary. *Multimed Tools and Applications*, 79, 26549–26569, 2020. https://doi.org/10.1007/s11042-020-09329-z
9. Debnath, Bikash, Das, Jadav Chandra, De, Debashis Reversible logic-based image steganography using quantum dot cellular automata for secure nano communication. *IET Circuits, Devices & Systems*, 11(1), 58–67, 2017.
10. Emad, E., Safey, A., Refaat, A. et al. A secure image steganography algorithm based on least significant bit and integer wavelet transform. *Journal of Systems Engineering and Electronics*, 29(3), 639–649, June 2018. https://doi.org/10.21629/JSEE.2018.03.21

11. Ghosal, S.K., Chatterjee, A., Sarkar, R. Image steganography based on Kirsch edge detection. *Multimedia Systems*, 27, 73–87, 2021. https://doi.org/10.1007/s00530-020-00703-3

12. Li, Y., Xiong, C., Han, X. et al. Retracted article: Image steganography using cosine transform with large-scale multimedia applications. *Multimed Tools and Applications*, 79, 9665, 2020. https://doi.org/10.1007/s11042-017-5557-1

13. Ghasemzadeh, Hamzeh, Kayvanrad, Mohammad H. Comprehensive review of audio steganalysis methods. *IET Signal Processing*, 12(6), 673–687, 2018.

14. Zarmehi, Nematollah, Akhaee, Mohammad Ali Digital video steganalysis toward spread spectrum data hiding. *IET Image Processing*, 10(1), 1–8, 2016.

15. Baagyere, E. Y., Agbedemnab, P. A., Qin, Z. et al. A multi-layered data encryption and decryption scheme based on genetic algorithm and residual numbers. *IEEE Access*, 8, 100438–100447, 2020. https://doi.org/10.1109/ACCESS.2020.2997838

16. Saad, A. H. S., Mohamed, M. S., Hafez, E. H. Coverless image steganography based on optical mark recognition and machine learning. *IEEE Access*. https://doi.org/10.1109/ACCESS.2021.3050737

17. Wu, Hanzhou, Wang, Wei, Dong, Jing et al. A graph-theoretic model to steganography on social networks. Available online at https://arxiv.org/abs/1712.03621, Last accessed: August 2020.

18. Liu, G., Liu, W., Dai, Y. et al. Adaptive steganography based on block complexity and matrix embedding. *Multimedia Systems*, 20(2), 227–238, 2014.

19. Xiang, L., Wu, W., Li, X. et al. A linguistic steganography based on word indexing compression and candidate selection. *Multimed Tools and Applications*, 77, 28969–28989, 2018. https://doi.org/10.1007/s11042-018-6072-8

20. Taleby Ahvanooey, M., Li, Q., Hou, J. et al. AITSteg: An innovative text steganography technique for hidden transmission of text message via social media. *IEEE Access*, 6, 65981–65995, 2018. https://doi.org/10.1109/ACCESS.2018.2866063

21. Wang, K., Gao, Q. A coverless plain text steganography based on character features. *IEEE Access*, 7, 95665–95676, 2019. https://doi.org/10.1109/ACCESS.2019.2929123

22. Wu, D., Su, H. Steganography via E-books with the EPUB format by rearrangements of the contents of the CSS files. *IEEE Access*, 8, 20459–20472, 2020. https://doi.org/10.1109/ACCESS.2020.2966889

23. Zhang, Z., Fu, G., Ni, R. et al. A generative method for steganography by cover synthesis with auxiliary semantics. *Tsinghua Science and Technology*, 25(4), 516–527, Aug. 2020. https://doi.org/10.26599/TST.2019.9010027

24. Kraetzer, C., Dittmann, J. Steganography by synthesis: Can commonplace image manipulations like face morphing create plausible steganographic channels?" In *Proceedings of the 13th International Conference on Availability, Reliability and Security, ARES 2018, Association for Computing Machinery*, New York, NY, USA, Article 11, 1–8. https://doi.org/10.1145/3230833.3233263

25. Qin, J, Wang, J, Tan, Y et al. Coverless image steganography based on generative adversarial network. *Mathematics*, 8(9), 1394, 2020. https://doi.org/10.3390/math8091394

26. Zhou, Z., Mu, Y., Wu, Q. M. Coverless image steganography using partial-duplicate image retrieval. *Soft Computing*, 23, 4927–4938, 2019.

27. Liao, X., Yu, Y., Li, B. et al. A new payload partition strategy in color image steganography. *IEEE Transactions on Circuits and Systems for Video Technology*, 30(3), 685–696, March 2020. https://doi.org/10.1109/TCSVT.2019.2896270

28. Liao, X., Yin, J., Chen, M. et al. Adaptive payload distribution in multiple images steganography based on image texture features. *IEEE Transactions on Dependable and Secure Computing*. https://doi.org/10.1109/TDSC.2020.3004708

4

Cyberattacks and Countermeasures in RPL-based Industrial Internet of Things

Cong Pu

Marshall University, Huntington, United States

CONTENTS

4.1 Introduction

As we enter the third decade of the 21st century, we are heralding the dawn of a new era of internet of things (IoT). The conception of IoT, a large number of intelligent objects seamlessly collaborating to realize the objectives [1], has spread around the world and became an after-dinner conversation for people. For example, leveraging IoT and existing state-of-the-art technologies (i.e., 5G, blockchain, and artificial intelligence [2]), industrial ecosystem is moving toward the time of Industry 4.0, which is widely known as industrial internet of things (IIoT) [3]. According to GlobeNewswire [4], the productivity benefits of IIoT are about to reach $265 billion in 2027. For example, according to the market survey report from Statista, the number of active IoT objects all over the world will reach 39 billion by 2025. As the usage of IIoT continues to grow across the world, there are high-demand jobs and employments in areas like data science, cybersecurity, and engineering and operations [5]. Data and information are at the core of IIoT. The true value of a ton of data and information generated by IIoT smart devices is to provide new insights for industry after being processed by various statistical and analytical techniques. As a substantial amount

DOI: 10.1201/9781003241348-4

of data and information are being stored and processed, protecting them from escalating cyberattacks becomes a challenging issue. For example, ransomware can target IIoT device hardware as well as applications and data. From July to September 2020, Check Point Research has reported that the daily average number of ransomware attacks increased by 50% compared with the number obtained from January to June 2020. To protect various assets of IIoT, security experts need to be well-prepared with basic and advanced cybersecurity technologies such as intrusion detection and prevention systems, threat modeling and security analysis, and attack detection and defense. As industries keep investing in various IIoT technologies, it is no wonder that the IIoT makes significant contribution to the progress of mankind.

To achieve the goals and realize the vision of IIoT, efficient communication and reliable information distribution in the IIoT have an important role to play, deservedly, routing protocols come into the spotlight. The main goal of a routing protocol is to specify how IIoT devices communicate with each other to distribute information, which makes all IIoT devices be able to select routes between them in the IIoT network. Over the past few years, routing protocols of IIoT have received burgeoning attention in industry as well as academia. A telling example is the recent revelation that Cisco and IETF collaborate to propose a new communication scheme for networks with resource constraints and lossy communication links [6], called RPL [7], to address the devices with resource constrains such as those in the IIoT. In RPL routing protocol, there are a group of routing attributes that can be regarded as either routing constraints or routing metrics. If the routing attribute is being used as a constraint, the attribute can prune IIoT devices and communication links from candidate paths that do not respect the constraint. On the other side, the routing attribute is able to determine the least cost path if being used as a metrics. In academia, the authors in [8], [9], [10], [11], and [12] proposed various communication protocols to improve the communication efficiency of IIoT network devices. For instance, in [13], the researchers designed a tree-based communication protocol for IoT environment, where the consumption of energy and the delay of communication could be reduced by adopting a mobile sink. In order to maintain the routes in the IoT network, two approaches are introduced. The first approach is designed based on the traditional geographic routing protocol so that the consumption of energy can be properly balanced. The second approach tries to use a small number of control packets to maintain the tree-based routing structure. Assuredly, the above facts are enough to prove that routing protocols are the cornerstone of IIoT and should be treated with respect.

When network engineers and academic researchers design or propose routing protocols, however, functionality is their major focus, whereas security is an afterthought. Consequently, many well-performed routing protocols shine in a nurturing environment (i.e., network simulation), but fail in the realistic environment because they did not consider the security and privacy issues in their design. For instance, RPL routing protocol has several intrinsic and charming characteristics such as automated configuration, dynamic reaction to the change of network structure, routing loop correction, and the availability of different network instances [7]. To be specific, the feature of automatic configuration will make the network be able to discover routing paths dynamically. In terms of loop detection and avoidance, RPL routing protocol has an ability to identify routing loops whenever the topology of network changes and is able to repair the routing loops. Basically, RPL is believed to be one of the promising candidate communication algorithms for the IIoT because it has many attractive features that make it easier to satisfy various industrial applications' quality of service (QoS) requirements. Although RPL routing protocol has sufficient maturity, several challenging problems still remain unsolved. For example, according to RPL specification, the implementation of various security features is partially or fully optional because of concerns about system performance. As a result, RPL routing protocol becomes vulnerable to several well-known attacks inherited from wireless network and RPL-specific attacks [14].

Although a large amount of effort has been dedicated to evaluating and enhancing RPL routing protocol's performance in various applications/systems, we will discuss the issues of security and privacy in RPL, mainly focusing on security attacks and countermeasures in the IIoT. This chapter is motivated in the matter of two facets. First, we concentrate on RPL routing protocol and IIoT, which are currently attracting a lot of attention because of their wide applicability. We carefully analyze the RPL routing protocol as well as the Trickle communication algorithm so that other researchers can obtain a better understanding about the RPL-based IIoT. Second, we select state-of-the-art RPL-specific attacks and analyze their malicious operations as well as performance impact on IIoT, which highlights the necessity of advanced defense mechanisms and countermeasures to protect RPL-based IIoT. In this chapter, RPL routing protocol and its major component, Trickle algorithm, are first introduced. Then, we identify and analyze various RPL-specific attacks in the IIoT and discuss their corresponding countermeasures. Finally, we conclude the chapter with future research directions, including interdisciplinary aspects and insights.

4.2 Related Work

Since RPL routing protocol was released, many academic researchers and industrial engineers have investigated the security and privacy issues of RPL and proposed state-of-the-art countermeasures to defend against various attacks.

In [15], the authors develop an intrusion detection system (also called DETONAR) to defend against RPL-specific attacks in the IoT. First, the authors conduct extensive experiments and collect network traffic for several RPL-specific attacks. The simulation results have been prepared and presented as a Routing Attacks Dataset for RPL (RADAR). Second, they develop an IDS, DETONAR, to detect some security attacks in RPL routing protocol. The main technique being utilized by DETONAR is packet sniffing, where a group of security policies (signature/anomaly-based rules) are adopted to detect suspicious activities from incoming Internet traffic. As reported by their experimental results, DETONAR's positive detection rate exceeds 80% for ten attacks with a small amount of computation overhead. The authors in [16] first investigate the version number attack which try to exhaust network resources (e.g., energy power, memory storage, and computation capability) by targeting the global repair mechanisms of RPL routing protocol. And then, they propose a version number attack detection mechanism. In the feature extraction method, a step forward feature selection scheme is used to choose the ideal features. Simulation and experiment results demonstrate that the detection mechanism is very competitive with good performance in detection accuracy and computation overhead. In [17], the authors conduct a literature review regarding RPL-related IDS in the IoT. Ref. [17]'s main contribution is that the authors identify several basic design requirements for intrusion detection systems based on various security attacks and their impacts. In addition, the authors discuss the best practices and research gaps in the research community of intrusion detection systems.

In [18], the authors put their efforts on a Destination-Oriented Directed Acyclic Graph (DODAG) Information Solicitation (DIS) attack, investigate the attack characteristics, and then design a defense mechanism in IoT network running with RPL. In the DIS attack, the adversary multicasts DIS messages to frequently reset the timer of DODAG Information Object (DIO) messages, resulting in control messages congestion in the network. To identify the DIS attack, the researchers propose a countermeasure which can reduce the response rate of DIO messages to DIS messages. The experimental study has proven that the communication overhead as well as energy consumption can be reduced. The authors in [19] propose a deep learning–based detection mechanism to detect/mitigate hello flooding attack in the IoT setting. The primary goal of hello flooding attack is to consume the limited resources of IoT devices. The authors conduct experiments with the comparison of existing benchmark schemes such as support vector machines (SVM) and logistic regression. The authors in [20] investigate sybil attack and propose a defense mechanism. The basic idea is that all nodes are organized into a tree structure and each non-leaf node stores a detection table in the memory. When receiving a packet, the node examines the piggybacked source node ID and previous hop node ID with the entries in the detection table. If there is no matched entry, the piggybacked source node ID and previous hop node ID are added in the detection table. If there is a matching entry in the detection table, the receiving node broadcasts an alarm packet to announce the suspected sybil attack. After careful analysis, it is found that the proposed scheme has a huge computation and storage overhead. For example, if there are a huge number of entries in the detection table, the sybil attack detection overhead will significantly increase. In [21], the authors identify a new RPL-specific attack, named non-spoofed copycat attack, and investigate its performance impact against IPv6-based wireless personal area networks. Through exploiting non-spoofed copycat attack, attackers implicitly monitor the DIO messages of neighbors, and then replay the captured DIO messages several times. The primary purpose of non-spoofed copycat attack is to affect RPL's performance (i.e., communication latency and packet delivery ratio [PDR]) in IPv6-based wireless personal area networks.

In [22], the authors adopt the trust technique to detect sybil attack, where the behaviors of devices are evaluated based on the current and old trust values. If a sybil attack is detected, the adversary will be assigned with a lower trust value. As a result, the nodes with a lower trust value (possibly the adversary) cannot participate in the regular routing operations. In [23], an intrusion detection system, named SVELTE, is proposed to detect network and routing layer attacks. In SVELTE, the intrusion detection scheme and firewall collaborate to examine network traffic and identify suspicious activities. In [24], the authors propose a camouflage-based approach to detect packet-dropping attack in wireless networks with energy harvesting capability. In the proposed approach, each node intentionally disguises to be energy harvesting node (i.e., cannot perform implicit monitoring), and then stealthily monitors neighbor node's forwarding operation. If the adjacent neighbor chooses not to re-send the received packet, packet-dropping attacks can be detected. In [25], the authors investigate a rank attack in the IoT network, where the adversary targets the rank value in RPL routing protocol and compromises the rank rule to affect the network performance. In [26], the authors propose a authentication scheme using secure hash functions to protect the communication from adversaries in the RPL-based IoT. The authors in [27] present an analysis of security threat for RPL. In their work, the potential security challenges and basic defense mechanisms are presented and analyzed. In [28], the authors focus on IEEE 802.15.4 medium access control protocol and its usage and limitation in the IoT environment. A survey about denial-of-service attacks against IoT is provided in

[29, 30]. The authors in [31] review the history and development of RPL routing protocol and point out several directions for future research.

4.3 Background

4.3.1 RPL Routing Protocol

The IIoT helps once "dump" devices get more intelligent by empowering them to collect, process, and send data traffic over the Internet, as well as communicate with other IIoT devices and information and communication systems seamlessly [32]. The acronym IIoT has a promise of improving the efficiency of regular operations with the assistance of artificial intelligence, machine learning, and advanced wired/wireless communication technologies. But, the IIoT encompasses a broad range of industrial-grade applications, and its application range is very wide. Far from being restricted to just the concept model, the IIoT can be found in a variety of domains, from ABB Smart Robotics, to Airbus Future Factory, to Amazon Smart Warehousing [33]. The IIoT devices deployed for these applications typically operate under capacity constraints in terms of processing and storage capability and battery energy [34]. The interconnection links between IIoT devices are featured with relatively low data rate and high packet loss ratio. In addition, the IIoT network can scale the number of devices from a few dozen to thousands, and the communication traffic can be classified into one-to-one, one-to-many, and many-to-one modes. To achieve efficient and reliable communications in an IIoT environment, the routing protocol for networks with resource constraints and lossy communication links [7], widely known as RPL, was proposed by Cisco together with IETF in 2012.

RPL-based IIoT example is shown in Figure 4.1, where there are three DODAGs and two RPL instances. Specifically, RPL routing protocol organizes IIoT devices (later nodes) into a hierarchical tree structure which is termed the Destination-Oriented Directed Acyclic Graph (DODAG). Typically, a DODAG contains a gateway node and a set of regular nodes. Here, the gateway node is termed the DODAG root, connecting with the Internet so that

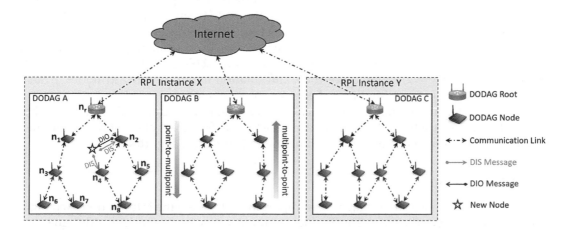

FIGURE 4.1
IIoT running with RPL, three DODAGs forming two instances of RPL [57].

IIoT nodes can communicate across several networks. For large-scale IIoT networks (e.g., thousands of nodes are deployed in the industrial environment), nodes can be self-organized into numerous DODAGs, and DODAGs can be further grouped into different RPL instances. For multiple DODAGs from the identical RPL instance, they are given the same RPL instance ID. The rationale of creating and maintaining multiple RPL instances is to arrange different tasks in the IIoT network, where one RPL instance is responsible for one task and operates independently from other RPL instances. For example, two RPL instances can be established in the IIoT network, where one is liable for collecting and transferring temperature data, and the other is accountable for monitoring the movements of people. Every node has a rank value. The more closer (or far) the node is located from the root of DODAG, the more smaller (or larger) the value of rank will be. In addition, the value of rank is being utilized to avoid DODAG loops as well as allow nodes to differentiate parent, child, and sibling nodes in the DODAG.

Speaking of traffic flows in RPL routing protocol, three different communication paradigms are supported: (i) one-to-one communication; (ii) one-to-many communication; and (iii) many-to-one communication. The one-to-one communication is adopted by any arbitrary pair of nodes in the DODAG to communicate. When the DODAG root has a command to be issued to other regular nodes in the DODAG, it employs the one-to-many communication mode. If the normal nodes have data for the DODAG root, they can use many-to-one communication mode.

In order to realize all the above-mentioned functionalities, four control messages are defined in RPL routing protocol: DIO, DAO, DAO-ACK, and DIS. DIO message is carrying network-relevant information. Usually, the root node of DODAG initiates DIO message to form a new DODAG, establish downward routing paths, and assist new nodes to discover nearby DODAG to join. DAO message is transmitted by the leaf nodes of DODAG so that the upward routing information can be propagated from the leaf nodes to the root node of DODAG. DAO-ACK message is utilized to confirm that DAO messages have been successfully received. New nodes will adopt DIS messages to join the existing DODAG in the network.

To maintain DODAG downward routes, RPL routing protocol can be configured to operate with either of two modes: (i) caching mode and (ii) non-caching mode. The basic idea of caching mode is that the route information to downwards nodes are stored by each node in the memory. For example, if a DAO message is received by a node, it first caches the piggybacked route and then adds its node identifier in the aggregated route and passes on the DAO message to the upward node. When the node receives a packet and the destination is one of its descendant nodes, it forwards the packet to the destination node via the cached downward route. If its descendant node is not the destination of packet, the packet will be sent to the parent node. However, in the non-caching mode, the root node of DODAG is the only node that stores route information about downward nodes. Thus, when there is a packet to send, the packet has to be sent via the upward route to the root node of DODAG. And then, the packet will be forwarded to the destination node via the cached downward route by the root node of DODAG. A flowchart of one-to-one communication with caching mode and non-caching mode is shown in Figure 4.2.

4.3.2 Trickle Algorithm

In an unstable and hash environment, it is important to regulate the information exchange robustly and energy efficiently. The Trickle communication algorithm [35] is designed to achieve the goal of information consistency in the network. The logic of Trickle communication algorithm is that a node is able to dynamically adjust its packet transmission ratio

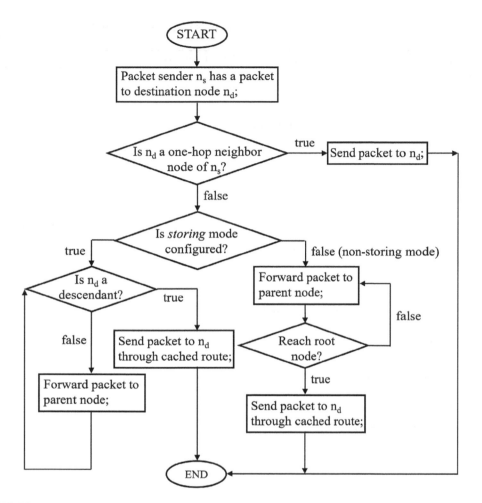

FIGURE 4.2
A flowchart of one-to-one communication with caching model and non-caching mode [57].

based on the degree of information consistency. RPL's Trickle communication algorithm is adopted to control the transmission rate of DIO messages. Since DIO packets usually carry critical network information, as a result, the transmission of DIO packets deserves special attention. To be specific, if an inconsistency is detected by a node, the node will increase the transmission rate of its DIO packets. On the other side, if there are no inconsistent information detected, the node will decrease its DIO packet transmission rate. According to [35], six major system parameters are adopted to achieve the goals of Trickle communication algorithm.

- I_{min}: the lower bound of timer.
- I_{max}: the upper bound of timer.
- k: the redundant parameter.
- I: the length of current timer.
- t: a time within the current timer.
- c: a counting parameter.

The pseudocode of controlling DIO packet transmissions is shown in Algorithm 1 [57].

Algorithm 1: Trickle Algorithm in RPL Routing Protocol

   Input: I_{min}, I_{max}, k
   Output: I, t, c

1 **Function** Init(I_{min}):
   /* sets timer I to the first interval   */
2 | $I \leftarrow I_{min}$;
3 | return I;

4 **Function** NewIntvl():
   /* doubles the interval length   */
5 | $I \leftarrow I \times 2$;
   /* sets a counter variable c to 0   */
6 | $c \leftarrow 0$;
   /* sets the interval length to I_{max} */
7 | if $I_{max} \leq I$ then
8 | | $I \leftarrow I_{max}$;
9 | end
   /* sets t to a random point in interval   */
10 | $t \leftarrow rand[\frac{I}{2}, I]$;
11 | return t;

12 **Function** RecConsTrans():
   /* increases counter variable c */
13 | $c \leftarrow c + 1$;
14 | return c;

15 **Function** RecConsTrans():
16 | if $I_{min} < I$ then
   | /* resets timer I */
17 | | $I \leftarrow I_{min}$;
18 | end
19 | return I;

20 **Function** TimerExp():
21 | if $c < k$ then
   | /* c less than redundancy constant k */
22 | | Transmit scheduled DIO;
23 | else
24 | | Suppress scheduled DIO;
25 | end

4.3.3 System and Adversary Models

We assume that the IIoT is composed of a set of DODAGs, where each node has constrained computing and storage capability and battery energy. In the DODAG, the DODAG root and regular nodes communicate directly and indirectly via unreliable links. In addition, each node is preassigned a unique ID [36]. Since the IIoT nodes are usually deployed in a wide-open or unattended area, they can be easily captured and compromised by the adversary [37]. Through probing attacks, the adversary might be able to access the security-critical module of integrated circuit, retrieve sensitive data, and then reprogram it to turn it into the malicious node [38].

The IIoT nodes usually need to keep operating for a period of time in the areas of interest [39]. If the IIoT nodes are equipped with regular batteries and dedicated in the operations of monitoring and communication, their battery power only can last for 5.8 days [40]. Consequently, replacing the battery of IIoT nodes becomes inevitable so that the IIoT network can survive and continue to operate. However, the IIoT nodes are usually deployed in difficult-to-reach locations, thus replacing or refilling batteries becomes very challenging or even impossible. Thus, people usually choose to use drones to deploy new IIoT nodes in the interest of area to maintain the operation of network [41].

4.4 RPL-Specific Attacks and Countermeasures

RPL is vulnerable to various cyberattacks because of the nature of wireless medium, the resource constraints of IIoT nodes, and the optional implementation of security mechanisms [42]. In addition, the security and privacy issues were not the first concerns when RPL was designed, thus it cannot satisfy the security requirements of current critical systems. It has been discovered that none of the existing IIoT operating systems [43] [44] include the implementation of RPL's security mechanisms. There are several existing security mechanisms to protect RPL-based IIoT from security attacks, such as intrusion detection systems [17] and hash chain authentication technique [45]. These security mechanisms can effectively detect outside attackers; however, they cannot defend against any inside attackers. An adversary might control a subset of IIoT nodes in the network and also has access to the cryptographic keys so that the cryptographic protection mechanisms can be easily bypassed. In the following, several representative RPL-specific attacks are presented and their corresponding countermeasure is briefly discussed.

4.4.1 Sybil Attack

For a new node, it can transmit a DIS message to request DODAG-related information from adjacent nodes to join the network. After receiving the DIS message, the neighbor node prepares and responds with a DIO message carrying RPL instance ID, version number, rank value, and other network configuration parameters. With those network-relevant information, the new node can choose a parent node, calculate the value of rank, and join the network. Unfortunately, the attackers can exploit the vulnerability of DIS messages to perform sybil attack against the IIoT network [46]. Specifically, the attackers first create many malicious DIS packets with fake node IDs and then broadcast those packets. Here, the fake node identifiers could be media access control addresses that are randomly generated by the adversary. If a normal node receives the malicious packet, it thinks that fresh nodes are eager to become members of network. Based on the Trickle communication scheme, the normal node needs to restart its DIO Trickle timeout period and then broadcast DIO packets. However, broadcasting DIO packets will require the normal node to consume energy resources [47].

For example, in Figure 4.3, an adversary A is launching sybil attack against the legitimate node n_1, n_2, and n_4 by broadcasting an attack DIS message with the fake identifier. It is clearly shown that n_1 and n_2 are important bridge nodes between the root node of DODAG and other nodes in the DODAG. If the bridge nodes are not available (i.e., consuming all energy resource), the network will be divided into several parts. When n_1 and

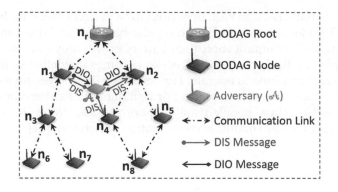

FIGURE 4.3
Sybil attack.

n_2 receive the DIS message, they assume that a non-member node might want to join the DODAG and is soliciting DODAG-related information from neighbor nodes. According to RPL specification, both n_1 and n_2 first restart their DIO Trickle timeout period to I_{min} and then broadcast a DIO message when the timer expires. Here, because n_1 and n_2 are not adjacent nodes, the DIO message from one node will not be able to prevent the DIO message from the other node. When the attacker A transmits a large number of malicious DIS messages with fictitious ID, n_1 and n_2 have to respond by transmitting the corresponding amount of DIO messages. Because of the frequent receiving and sending of messages, n_1 and n_2 will quickly consume their limited power of batteries. As a result, the lifetime of nodes will be shortened. When the battery energy is completely exhausted, the network partition will be formed in the DODAG, where other nodes (e.g., n_3) will not be able to communicate with the DODAG root n_r anymore. To demonstrate the severe consequences of sybil attack in the DODAG as shown in Figure 4.3, a preliminary experiment is conducted in OMNeT++ [48]. According to [35], we set $I_{min} = 0.1$ sec, $I_{max} = 6,554$ sec, and $k = 1$ in the experiment. In Figure 4.4, we obtain the amount of DIO packets against the experimental time. According to Figure 4.4, it is clearly shown that the amount of transmitted

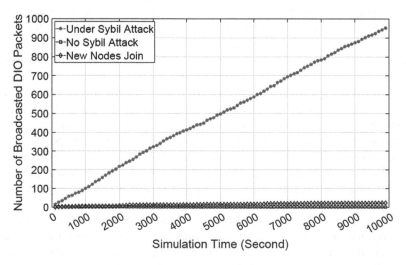

FIGURE 4.4
The impact of sybil attack [57].

DIO messages increases when the length of experimental time increases. On the other side, for the scenario without adversary, the number of transmitted DIO messages is maintained at the relatively low level.

In [49, 50], a Gini coefficient technique is being adopted to detect sybil attack by measuring the statistical dispersion. To be specific, during the observation window, every network node keeps track of the statistical dispersion of node IDs in the DIS messages and then computes the Gini coefficient. After comparing the Gini coefficient with a predetermined boundary value, the legitimate node can detect the existence of sybil attacks. However, the proposed Gini coefficient detection scheme has one potential drawback: the detection latency and accuracy are closely related to the length of observation window. If the observation window is short, we can expect a low detection latency as well as a low detection accuracy. However, if the observation window is long, the detection latency will be increased, and we will have a high detection accuracy.

4.4.2 Packet-Dropping Attack

In RPL routing protocol, the DODAG root needs to issue a DIO packet to form a DODAG after the network nodes are deployed in an area of interest. The DIO message is typically piggybacked with the ID of root node of DODAG, the DODAG root node's rank value, as well as an objective function. Here, the objective function specifies how each node calculates the rank value based on predefined metrics. After receiving the message of DIO, if the network node plans to join the existing DODAG, the sender ID of DIO message can be added to the list of parent nodes. In addition, it computes the value of rank for itself and then passes on the DIO packet piggybacked with its own identifier and rank value to other nodes. If there are multiple nodes in a node's parent list, the member of parent list having the smallest value of rank becomes the preferred parent node. When the network node has data traffic to the root node of DODAG, the preferred parent node is automatically selected as the next-hop node to send data traffic. However, an adversary can intentionally put a smaller value of rank in the DIO packet and transmit the packet to other network nodes, which make other nodes select the adversary as their preferred parent node. So, when the adversary receives data traffic from other nodes, it can drop any data packets intentionally to deafen the root node of DODAG.

For example in Figure 4.5, an adversary A is launching packet-dropping attack in the DODAG. First, the adversary A broadcasts a DIO message with a smaller value of rank to

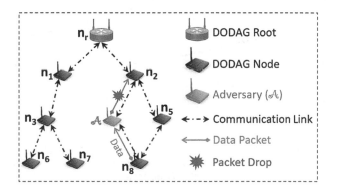

FIGURE 4.5
Packet-dropping attack [57].

make its child node (e.g., n_8) to choose it as the preferred parent node. Then, when n_8 sends a data packet to the adversary A, the adversary A randomly or strategically drops the data traffic without forwarding them to the next-relay node (e.g., n_2). As a result, the DODAG root n_r cannot receive the data packet from n_8, and the network performance is significantly affected. We conduct a preliminary experiment to expose the severe consequences of packet-dropping attack in Figure 4.6, where the PDR is obtained with changing packet drop rate and channel error rate (r_{ch_err}). As shown in Figure 4.6, when the packet drop rate increases, the PDR significantly decreases. Because more data packets are being dropped by the adversary A, less number of data packets can be transmitted to the root node of DODAG and a smaller PDR is obtained. In addition, the PDR is significantly affected by the channel error rate r_{ch_err}. The overall PDR decreases as the r_{ch_err} increases. The reason is that data packets might get missing in the process of forwarding because of the bad wireless channel quality. Consequently, a lower PDR is measured.

In [51], a monitor-based mechanism is designed to detect/mitigate malicious packet-dropping behaviors and isolate the adversary from the RPL-based IIoT. After sending the data traffic to its preferred parent node, the packet sender continues to monitor the follow-up operation of packet receiver. If the packet receiver drops the packet without forwarding, the packet sender can compare the ratio of packet loss of preferred parent node with the average packet loss rate of adjacent nodes. If the ratio of packet loss of preferred parent node is lower or smaller than the average packet loss rate of adjacent nodes, the packet-dropping activities of adversary can be detected. When the total number of caught packet-dropping activities is larger than a predefined boundary value, the detection node will issue an alarm packet so that all neighbor nodes will not send any packet to the adversary any more. The idea of monitor-based mechanism is straightforward and easy to implement; however, its disadvantage is obvious, too. The predefined threshold value should be carefully selected. If a larger threshold value is adopted, a higher miss detection rate might be observed. On the other side, if the threshold value is too small, a false positive rate will be high.

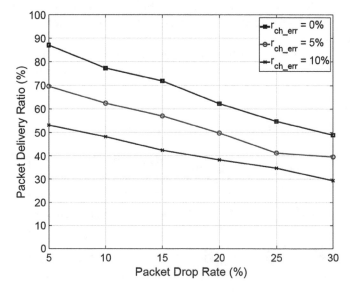

FIGURE 4.6
The impact of packet-dropping attack [51].

4.4.3 DAO Divergence Attack

In RPL routing protocol, the caching mode enables network nodes to actively obtain the downward routing paths to its descendant nodes through buffering the route information piggybacked in the DAO messages. Through caching the downward route information, every network node has the information about the next-relay node via which the data traffic can be delivered to the destination. However, the downward route in the buffer might become stale (e.g., one intermediate node along the route is not available), and the data traffic cannot be delivered to the destination if the stale route is selected to send data traffic. Thus, to get rid of stale routes from the buffer, the flag of Forwarding-Error in the header of the packet is utilized to quickly identify the unreachable next-hop node and report the error route in the network. To be specific, when a node is unable to send the network traffic to the next-relay node according to the cached routing information, it sets the Forwarding-Error flag, creates an error message, and then sends the error message to the parent node. When the error message with the Forwarding-Error flag set reaches the parent node, the parent node has to remove the reported downward route from its buffer space according to RPL specification. Originally, the Forwarding-Error flag is designed to report and discard the error route in the network and finally improve the performance of RPL routing protocol. However, this feature might be exploited by attackers to launch DAO divergence attack [52] against the IIoT systems.

In Figure 4.7, an adversary A receives a data packet from node n_2. Instead of forwarding the data traffic to the next-relay node n_8, the adversary A chooses to drop the data packet. After that, the adversary A creates an error packet with the flag of Forwarding-Error set and responds n_2 with the error packet. When n_2 receives the error packet, it believes that the cached downward route to n_8 is no longer available, thus it removes the cached downward route from the buffer. To show the severe consequences of DAO divergence attack in the IIoT network, we conduct a preliminary experiment. A downward route which consists of 11 nodes is set up in the network, where node n_1 and n_6 are the DODAG root and the adversary A, respectively. Here, the attack rate indicates how frequently the adversary n_6 replies the error packets with the flag of Forwarding-Error set. Each node's energy usage along the downward route is measured in Figure 4.8, where the energy consumption of each node (i.e., n_1, n_2, n_3, n_4, and n_5) located prior to the adversary n_6 goes up when attackers perform more attacks. As the adversary n_6 generates and replies more error packets, its parent node has to discard valid downward routing information cached within the buffer. If its parent or ancestral node has network traffic for the same destination later, the parent

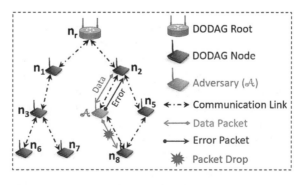

FIGURE 4.7
DAO inconsistency attack [57].

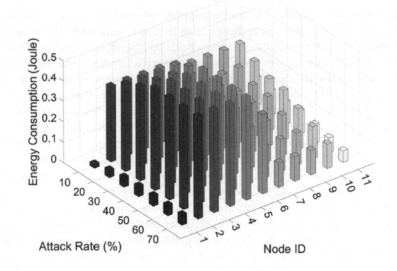

FIGURE 4.8
The consequences of DAO divergence attack [52].

or ancestral node needs to generate and reply an error packet to their corresponding parent node. This is because the downward routing information which can be used to reach the destination nodes has already been discarded. Consequently, the adversary A's ancestors along the downward route need to respond with error packets, which makes the energy consumption of intermediate nodes increase. To defend against DAO inconsistency attack, the parent node can set up a dynamic error packet acceptance rate which is adaptively fine-tuned according to the frequency of receiving error packets and the estimated channel error rate. With the error packet acceptance rate, each node can prevent the valid downward routes from being discarded.

4.4.4 Hatchetman Attack

In Hatchetman attack, when the adversary A receives the data packets, it first creates a huge amount of invalid packets by adding the error route information and then sends these attack packets to normal nodes, which can make normal nodes discard the received attack packets and answer with a lot of error messages to the root node of DODAG [53]. In consequence, the legitimate nodes not only drop many data messages but also reply many error messages, which waste the limited energy and communication resources. For example, in Figure 4.9, the DODAG root n_r generates a packet piggybacked with the route ([r, 1, A, 2, 3, 4, 5]) and sends it to node n_5. When the data packet reaches the attacker A, the attacker A first replaces the post-hop nodes (i.e., 3, 4, and 5) of the target node (i.e., n_2) with a fake ID of destination node (i.e., n_f). After that, the attacking packet carrying the error routing information ([r, 1, A, 2, f]) is forwarded to the node n_2. In the error route, n_f is the fictitious destination node address which is unreachable. When n_2 receives the packet with the unreachable route, it attempts to send the packet to the node n_f. However, n_f does not exist in the network and the forwarding operation fails. After that, n_2 needs to drop the received packet and issue an error message to the root node of DODAG n_r. If the adversary A transmits the false packet piggybacked with the unreachable routing information to each

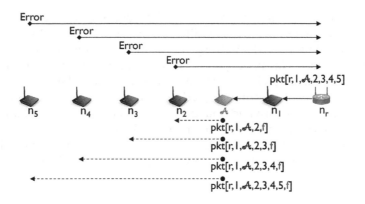

FIGURE 4.9
Hatchetman attack [53].

downstream node, all received false packets will be dropped and an error packet will be replied by each downstream node. It is shown in Figure 4.9 that the attacker A transmits the attack packets with the unreachable route to its downstream node, e.g., n_2, n_3, n_4, and n_5. After receiving the packet with the error route, all downstream nodes need to discard the packet. The reason is that the next-relay node is not reachable. Moreover, the downstream nodes have to generate and send error packets to the DODAG root n_r. In such case, every node between the attacker and the destination node has to receive and transmit many error packets, which consumes non-negligible amount of energy and communication resources in the network.

In Figure 4.10, the ratio of packet delivery (PDR) is obtained with changing channel error rate (r_{cer}) and the percentage of attackers (r_{ap}). When r_{cer} is 0%, the highest PDR is obtained. Because each node collaboratively forwards the received packets, the destination

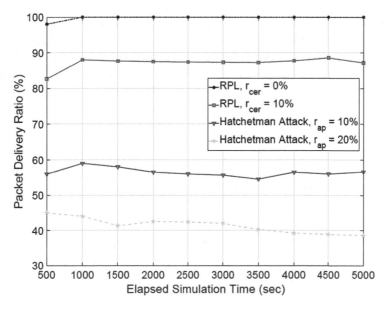

FIGURE 4.10
The impact of Hatchetman attack [53].

node will receive more packets and the highest PDR is shown. Without Hatchetman attack, RPL's performance is vulnerable to the quality of wireless medium, where the PDR is oscillating around 76% with r_{cer} = 10%. With r_{ap} = 10% and 20%, the smallest PDR is observed by the Hatchetman attack, comparing to the scenario running traditional RPL routing protocol without Hatchetman attack. Because the attacker creates a large amount of attacking packets with the unreachable destination node and transmit them to the normal nodes, the normal nodes will drop more invalid data packets and a lower PDR is obtained. When the number of adversaries performing Hatchetman attack increases, r_{ap} = 20%, the PDR drops below 45%. This is because more attackers can create more attack packets with the unreachable route. So, when the normal nodes receive those attack packets, they have to drop them.

4.4.5 Energy Abusing Attack

For RPL, the one-to-one communication is designed for arbitrary pair of DODAG nodes to communicate [54]. If the packet sender has a packet to its adjacent neighbor node, it just forwards the packet to the adjacent node directly, rather than sending the packet to its preferred parent node. In all other cases, the configuration of RPL, either caching mode or non-caching mode, will determine how the one-to-one communication is executed. To be specific, if the non-storing mode is configured in RPL routing protocol, the nodes, except for the DODAG root, do not cache any downward route toward descendant nodes. If there are some packets to send, the network node first has to transmit the packet to the root node of DODAG via the upward routing path. After receiving the packet, the DODAG root attaches the source routing information and transmits the packet to the destination node. However, if the RPL routing protocol is configured with storing mode, then each node will store the downward route toward descendant nodes. If there is a packet to its descendant, the packet can be forwarded via the cached downward route and finally reaches the descendant node. Otherwise, the packet is supposed to be sent to the root node of DODAG via sender's preferred parent node. After the packet reaches the root node of DODAG, the root node of DODAG will then send the packet to the destination node.

In the RPL-based IIoT, the one-to-one communication is often used for sending data traffic as well as end-to-end acknowledgments between arbitrary pair of nodes in the DODAG. Unfortunately, the one-to-one communication could be exploited by the adversary in the malicious manner to affect the performance of network. In Figure 4.11, the

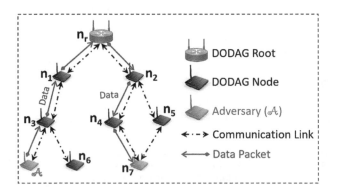

FIGURE 4.11
Energy abusing attack [55].

adversary A transmits many data packets to node n_7. If RPL routing protocol is configured with non-caching mode, the data packets have to be first transmitted to the root node of DODAG n_r via the upward routing path. Then, the root node of DODAG n_r attaches the source routing information in the data packets and sends them to n_7. If the caching mode is being configured in RPL routing protocol, the data packets need to be forwarded to the first common ancestor of adversary A and node n_7, which is the root node of DODAG n_r. After that, the DODAG root n_r forwards all data packets to n_7. According to the above analysis, it does not matter which mode RPL routing protocol is currently using, a sequence of intermediate nodes, such as n_3, n_1, n_r, n_2, and n_4, need to send a large number of data packets. However, those many sending operations will cause an increase in energy consumption. Because each IIoT node has very limited battery power, the energy depletion attack can quickly exhaust each node's battery power and finally causes the network unable to work.

In Figure 4.12, the energy usage is obtained with varying number of attackers and the attack rate (r_{atk}). Here, the attack rate r_{atk} indicates the frequency of sending attack data packets by the adversary. It is shown in Figure 4.12 that RPL routing protocol achieves the lowest energy consumption if there is no adversary in the network. As there are more attackers existing in the network, an increasing energy consumption can be observed under energy abusing attack (EDA). Because more attack data packets are being generated by the attackers, each intermediate node will need to transmit more attack packets. As a result, the energy consumption will increase. In addition, the energy consumption increases when attackers perform more attacks. With a larger attack rate r_{atk}, attackers will send more attack data packets. As a result, more energy resource has to be consumed by intermediate nodes because of frequent receiving and forwarding operations.

In [55], the authors propose a detection mechanism to detect/mitigate the EDA based on the nodes' behavior in RPL-based IIoT. The logic of the detection mechanism is that the number of sent packets during a predefined time period is being recorded by the adjacent

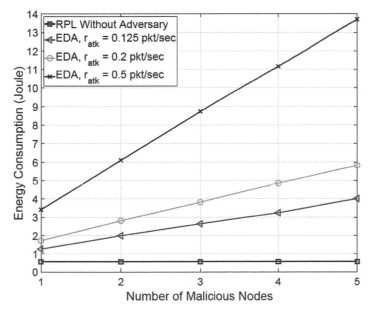

FIGURE 4.12
The impact of energy abusing attack [55].

node. Then, the counting number will be compared with a threshold value. When the number of sent packets is larger than the boundary value, the EDA can be detected.

4.5 Conclusion and Future Research Directions

The IIoT consisting of seamlessly interconnected smart devices has been seen in various industrial domains such as automated monitoring of inventory, quality control, supply chain optimization, and plant safety improvement. To improve the communication efficiency and revolutionize traditional industrial processes, a routing protocol named RPL has been proposed for the IIoT. Because the deployment of security mechanisms is missing in RPL routing protocol, the IIoT is vulnerable to several well-known attacks inherited from wireless network and RPL-specific attacks. This chapter introduces several representative RPL-specific cyberattacks, such as sybil attack, packet-dropping attack, DAO divergence attack, Hatchetman attack, and EDA, and discusses the potential countermeasure against these attacks in the IIoT.

To further explore the potential of IIoT, we recommend two promising research domains with the trans-disciplinary opinions for future investigation. First, because IIoT devices are usually powered by traditional battery, as a result, battery replacement or energy replenishment is unavoidable or even impossible [40]. Thus, energy harvesting becomes an ideal solution to extend the lifetime of IIoT devices [24]. Second, the traditional cryptographic schemes can provide fundamental protections, i.e., confidentiality, integrity, and availability. Unfortunately, those cryptographic schemes cannot be directly applied to IIoT devices that operate with resource-limited microprocessors [56]. Thus, the lightweight security protocols become the only solution to secure IIoT networks.

Finally, we believe that this chapter has the potential to help researchers discover novel research directions to pursue and contribute to the IIoT community through providing the detailed discussion of cyberattacks, countermeasures, and future research directions.

References

1. K. Tange, M. D. Donno, X. Fafoutis, and N. Dragoni, "A Systematic Survey of Industrial Internet of Things Security: Requirements and Fog Computing Opportunities," *IEEE Communications Surveys & Tutorials*, vol. 22, no. 4, pp. 2489–2520, 2020.
2. C. Pu, "A Novel Blockchain-Based Trust Management Scheme for Vehicular Networks," in *IEEE Proc. WTS*, pp. 1–6, 2021.
3. M. Bansal, A. Goyal, and A. Choudhary, "Industrial Internet of Things (IIoT): A Vivid Perspective," *Inventive Systems and Control*, vol. 204, pp. 939–949, 2021.
4. Industrial IoT (IIoT) Market Worth $263.4 Billion by 2027, 2020. https://www.globenewswire.com/
5. The Opportunities of the Industrial Internet of Things. https://www.roevin.ca/blog/2019/august/opportunities-industrial-internet-of-things/
6. Securing the Internet of Things: A Proposed Framework. https://tools.cisco.com/security/center

7. T. Winter, et al., "RPL: IPv6 Routing Protocol for Low-Power and Lossy Networks," *RFC*, vol. 6550, pp. 1–157, 2012.
8. F. Al-Turjman, "Cognitive Routing Protocol for Disaster-Inspired Internet of Things," *Future Generation Computer Systems*, vol. 92, pp. 1103–1115, 2019.
9. T. Behera, S. Mohapatra, U. Samal, M. Khan, M. Daneshmand, and A. Gandomi, "I-SEP: An Improved Routing Protocol for Heterogeneous WSN for IoT-Based Environmental Monitoring," *IEEE Internet of Things Journal*, vol. 7, no. 1, pp. 710–717, 2019.
10. R. Coutinho, A. Boukerche, and A. Loureiro, "A Novel Opportunistic Power Controlled Routing Protocol for Internet of Underwater Things," *Computer Communications*, vol. 150, pp. 72–82, 2020.
11. B. Djamaa, M. Senouci, H. Bessas, B. Dahmane, and A. Mellouk, "Efficient and Stateless P2P Routing Mechanisms for the Internet of Things," *IEEE Internet of Things Journal*, vol. 8, no. 14, pp. 11400–11414, 2021.
12. S. Jazebi and A. Ghaffari, "RISA: Routing Scheme for Internet of Things using Shufled Frog Leaping Optimization Algorithm," *Journal of Ambient Intelligence and Humanized Computing*, vol. 11, no. 10, pp. 4273–4283, 2020.
13. R. Yarinezhad and S. Azizi, "An Energy-Efficient Routing Protocol for the Internet of Things Networks Based on Geographical Location and Link Quality," *Computer Networks*, vol. 193, p. 108116, 2021.
14. A. Verma and V. Ranga, "Security of RPL Based 6LoWPAN Networks in the Internet of Things: A Review," *IEEE Sensors Journal*, vol. 20, no. 11, pp. 5666–5690, 2020.
15. A. Agiollo, M. Conti, P. Kaliyar, T. Lin, and L. Pajola, "DETONAR: Detection of Routing Attacks in RPL-Based IoT," *IEEE Transactions on Network and Service Management*, vol. 18, no. 2, pp. 1178–1190, 2021.
16. M. Osman, J. He, F. Mokbal, N. Zhu, and S. Qureshi, "ML-LGBM: A Machine Learning Model based on Light Gradient Boosting Machine for the Detection of Version Number Attacks in RPL-Based Networks," *IEEE Access*, vol. 9, pp. 83 654–83 665, 2021.
17. G. Simoglou, G. Violettas, S. Petridou, and L. Mamatas, "Intrusion Detection Systems for RPL Security: A Comparative Analysis," *Computers & Security*, vol. 104, p. 102219, 2021.
18. F. Medjek, D. Tandjaoui, N. Djedjig, and I. Romdhani, "Multicast DIS Attack Mitigation in RPL-based IoT-LLNs," *Journal of Information Security and Applications*, vol. 61, p. 102939, 2021.
19. S. Cakir, S. Toklu, and N. Yalcin, "RPL Attack Detection and Prevention in the Inter-net of Things Networks Using a GRU Based Deep Learning," *IEEE Access*, vol. 8, pp. 183 678–183 689, 2020.
20. P. Kaliyar, W. Jaballah, M. Conti, and C. Lal, "LiDL: Localization with Early Detection of Sybil and Wormhole Attacks in IoT Networks," *Computers & Security*, vol. 94, p. 101849, 2020.
21. A. Verma and V. Ranga, "CoSec-RPL: Detection of Copycat Attacks in RPL based 6LoW-PANs using Outlier Analysis," *Telecommunication Systems*, vol. 75, pp. 43–61, 2020.
22. D. Airehrour, J. Gutierrez, and S. Ray, "SecTrust-RPL: A Secure Trust-Aware RPL Routing Protocol for Internet of Things," *Future Generation Computer Systems*, vol. 93, pp. 860–876, 2019.
23. S. Raza, L. Wallgren, and T. Voigt, "SVELTE: Real-Time Intrusion Detection in the Internet of Things," *Ad Hoc Networks*, vol. 11, no. 8, pp. 2661–2674, 2013.
24. C. Pu and S. Lim, "Spy vs. Spy: Camouflage-based Active Detection in Energy Harvesting Motivated Networks," in *IEEE Proc. MILCOM*, pp. 903–908, 2015.
25. A. Le, J. Loo, A. Lasebae, A. Vinel, Y. Chen, and M. Chai, "The Impact of Rank Attack on Network Topology of Routing Protocol for Low-Power and Lossy Networks," *IEEE Sensors Journal*, vol. 13, no. 10, pp. 3685–3692, 2013.
26. A. Dvir, T. Holczer, and L. Buttyan, "VeRA-Version Number and Rank Authentication in RPL," in *Proc. IEEE MASS*, pp. 709–714, 2011.
27. T. Tsao, R. Alexander, M. Dohler, V. Daza, A. Lozano, and M. Richardson, "A Security Threat Analysis for the Routing Protocol for Low-Power and Lossy Networks (RPLs)," *RFC Standard* 7416, January 2015. Doi: 10.17487/RFC7416

28. S. M. Sajjad and M. Yousaf, "Security Analysis of IEEE 802.15. 4 MAC in the Context of Internet of Things (IoT)," in *Proc. IEEE CIACS*, pp. 9–14, 2014.
29. P. Kasinathan, C. Pastrone, M. A. Spirito, and M. Vinkovits, "Denial-of-Service Detection in 6LoWPAN Based Internet of Things," in *Proc. IEEE WiMob*, pp. 600–607, 2013.
30. A. Rghioui, A. Khannous, and M. Bouhorma, "Denial-of-Service Attacks on 6LoWPAN-RPL Networks: Threats and an Intrusion Detection System Proposition," *Journal of Advanced Computer Science & Technology*, vol. 3, no. 2, pp. 143–152, 2014.
31. H. Kim, J. Ko, D. Culler, and J. Paek, "Challenging the IPv6 Routing Protocol for Low-Power and Lossy Networks (RPL): A Survey," *IEEE Communications Surveys & Tutorials*, vol. 19, no. 4, pp. 2502–2525, September 2017.
32. C. Pu, J. Brown, and L. Carpenter, "A Theil Index-Based Countermeasure Against Advanced Vampire Attack in Internet of Things," in *IEEE Proc. HPSR*, pp. 1–6, 2020.
33. The Top 20 Industrial IoT Applications. https://www.iotworldtoday.com/2017/09/20/top-20-industrial-iot-applications/
34. K. Pister, P. Thubert, S. Dwars, and T. Phinney, "Industrial Routing Requirements in Low-Power and Lossy Networks," *RFC*, vol. 5673, pp. 1–27, 2009.
35. P. Levis, T. Clausen, J. Hui, O. Gnawali, and J. Ko, "The Trickle Algorithm," *Internet Engineering Task Force*, vol. RFC6206, pp. 1–13, 2011.
36. C. Pu and X. Zhou, "Suppression Attack Against Multicast Protocol in Low Power and Lossy Networks: Analysis and Defenses," *Sensors*, vol. 18, no. 10, p. 3236, 2018.
37. C. Pu, X. Zhou, and S. Lim, "Mitigating Suppression Attack in Multicast Protocol for Low Power and Lossy Networks," in *IEEE Proc. LCN*, pp. 251–254, 2018.
38. C. Pu, S. Lim, B. Jung, and J. Chae, "EYES: Mitigating Forwarding Misbehavior in Energy Harvesting Motivated Networks," *Computer Communications*, vol. 124, pp. 17–30, 2018.
39. C. Pu, S. Lim, B. Jung, and M. Min, "Mitigating Stealthy Collision Attack in Energy Harvesting Motivated Networks," in *IEEE Proc. MILCOM*, pp. 539–544, 2017.
40. C. Pu, T. Gade, S. Lim, M. Min, and W. Wang, "Lightweight Forwarding Protocols in Energy Harvesting Wireless Sensor Networks," in *Proc. IEEE MILCOM*, pp. 1053–1059, 2014.
41. S. Mnasri, N. Nasri, and T. Val, "The Deployment in the Wireless Sensor Networks: Methodologies, Recent Works and Applications," in *Proc. PEMWN*, 2014.
42. C. Pu and L. Carpenter, "Digital Signature Based Countermeasure Against Puppet Attack in the Internet of Things," in *IEEE Proc. NCA*, pp. 1–4, 2019.
43. Contiki Operating System. http://www.contiki-os.org/
44. TinyOS. http://www.tinyos.net/
45. G. Glissa, A. Rachedi, and A. Meddeb, "A Secure Routing Protocol based on RPL for Internet of Things," in *IEEE Proc. GLOBECOM*, pp. 1–7, 2016.
46. C. Pu, "Spam DIS Attack Against Routing Protocol in the Internet of Things," in *IEEE Proc. ICNC*, pp. 73–77, 2019.
47. C. Pu and S. Lim, "A Light-Weight Countermeasure to Forwarding Misbehavior in Wirless Sensor Networks: Design, Analysis, and Evaluation," *IEEE Systems Journal*, vol. 12, no. 1, pp. 834–842, 2018.
48. A. Varga, OMNeT++, 2014. http://www.omnetpp.org/
49. B. Groves and C. Pu, "A Gini Index-Based Countermeasure Against Sybil Attack in the Internet of Things," in *Proc. IEEE MILCOM*, pp. 1–6, 2019.
50. C. Pu, "Sybil Attack in RPL-Based Internet of Things: Analysis and Defenses," *IEEE Internet of Things Journal*, vol. 7, no. 6, pp. 4937–4949, 2020.
51. C. Pu and S. Hajjar, "Mitigating Forwarding Misbehaviors in RPL-based Low Power and Lossy Networks," in *IEEE Proc. CCNC*, pp. 1–6, 2018.
52. C. Pu, "Mitigating DAO Inconsistency Attack in RPL-based Low Power and Lossy Net- Works," in *IEEE Proc. CCWC*, 2018, pp. 570–574.
53. C. Pu and T. Song, "Hatchetman Attack: A Denial of Service Attack Against Routing in Low Power and Lossy Networks," in *IEEE Proc. CSCloud*, pp. 12–17, 2018.

54. C. Pu, "Energy Depletion Attack Against Routing Protocol in the Internet of Things," in *IEEE Proc. CCNC*, pp. 1–4, 2019.

55. C. Pu and B. Groves, "Energy Depletion Attack in Low Power and Lossy Networks: Analysis and Defenses," in *IEEE Proc. ICDIS*, pp. 14–21, 2019.

56. C. Pu and Y. Li, "Lightweight Authentication Protocol for Unmanned Aerial Vehicles Using Physical Unclonable Function and Chaotic System," in *IEEE Proc. LANMAN*, pp. 1–6, 2020.

57. C. Pu and K. Choo, "Lightweight Sybil Attack Detection in IoT Based on Bloom Filter and Physical Unclonable Function," *Elsevier Computers & Security*, vol. 113, p. 102541, 2022.

5

Cyber-physical Attacks and IoT

Mehtab Alam and Ihtiram Raza Khan

Jamia Hamdard, New Delhi, India

CONTENTS

DOI: 10.1201/9781003241348-5

5.1 Introduction

A cyber-attack is an intentional and malicious venture by an organization or an individual to gain unauthorized access to a user's or organization's databases or information systems. A cyber-attack can have numerous effects ranging from stealing of data, disabling of computer systems or entire networks, and a breached or hacked computer can be used at a later date as a launch point of bigger cyber-attacks. In today's world, cyber-attacks hit businesses every day. The Cisco Annual Cybersecurity Report has mentioned that the total number of cyber-attacks and related crimes has jumped up more than a hundred-fold between January and October 2019 (CISCO, 2019). Cybercrime is increasing on a daily basis as organizations and end users look to benefit from internet of things (IoT) and other such technologies. These malicious users, most of the times, are looking for ransom. Some attackers try to alter systems and data as a resemblance to "hacktivism" (Stefano Baldi, 2003).

In 2018, You et al. discussed a light weight behavior rule specifications-based monitoring solution in order to identify attacks on an embedded IoT device in a cyber-physical system (CPS; You, et al., 2018). In 2019, Sen et al. studied and analyzed cyber-attacks on big data IoT and CPS. They further discussed how to stabilize the various CPS (Sen & Jayawardena, 2019). In another article, Li et al. attempted to detect IoT cyber-physical attacks (CPA) based on energy auditing. With the help of energy meter readings, they developed a dual deep learning model system. They identified both cyber as well as physical attacks (Li, Shi, Shinde, Ye, & Song, 2019). In 2020, Rana et al. proposed a CPS that used IoT as the base communication system. They used a synchronous machine state estimation technique in their work (Rana, Khan, & Abdelhadi, 2020). In 2021, Elsisi et al. introduced an IoT architecture for monitoring gas-insulated switch-gear. It was derived from the concept of CPS in Industry 4.0 and CPA (Elsisi, et al., 2021). In another article, Jahromi et al. presented a two-level ensemble attack detection and attribution framework for CPS to prevent CPA (Jahromi, Karimipour, Dehghantanha, & Choo, 2021).

This chapter is organized into five sections. Section 5.1 is the introduction to the chapter. In Section 5.2, we have discussed cybercrimes and CPA. We have pointed some common types of cyber-attacks. Furthermore, a few cyber-attack trends and a few common types of CPA have been discussed. The next sub-section is about cybersecurity. In Section 5.3, we have given a detailed overview of IoT, how IoT works, the importance of IoT, benefits, pro, and cons of IoT and a few standards and frameworks of the IoT platform. The next subsection is about the security and privacy issues in IoT end-user devices. In Section 5.4, we have tried to bring out the effect of CPA on IoT. In Section 5.5, we have discussed a few methods to avoid CPA in IoT and some ways to prevent if not stop a cyber-attack. Some end-user protection methods have been highlighted followed by a few cyber safety tips. In Section 5.6, we have concluded the chapter. References are listed at the end.

5.2 Cybercrimes and Cyber-physical Attacks

The rapid urbanization worldwide, and the mass proliferation of the internet and internet-provided services and the fact that most of the biggest world economies and their critical infrastructure depend drastically on internet to function on a daily basis, has resulted in a number of upsides and as many downsides in a broader context (Hendrik & Mochalski, 2009). Upsides are clearly visible to everyone, be them a user, developer, worker, official etc. The internet is serving as one of the greatest inventions in the history of mankind (Van Deursen & Heisper, 2017). The digital transformation which took place the last few years has helped people to connect to everyone and anyone in any corner of the world. Above all, all of this with great and unprecedented easy, efficiency, creativity, innovation, and communication (Amour, 2012).

We have become completely accustomed to this everyday tool in our lives. The downside of this technology and interconnected systems and data repositories is the emerging dark side of the tech which is filled with cybercrimes, cyber-attacks, and malicious actors who are eyeing to make profits and steal vital information pushing the innocent people and business into jeopardy (Hampton, 2010). These victims have no option other than getting blackmailed and paying huge ransoms to keep themselves safe and protected. In most of the scenarios, either the data are lost forever or is in unreadable format.

Cyber-attacks are nowadays becoming the new nationwide or in some cases worldwide weapon, being used to threaten countries and their governments (Denning & Denning, 2010). Cybercrime is starting to dictate the global risk scenario at the top levels. It has broken all the barriers of the virtual world and has entered the physical world giving rise to the cyber-physical crimes and attacks (Anderson, et al., 2013).

5.2.1 Common Types of Cyber-attacks

In this section, we have listed a few types of cyber-attacks being launched to steal data and sensitive information. Figure 5.1 displays some common cyber-threats.

5.2.1.1 Malware

Any malicious software, including worms, viruses, spyware, and ransomwares, are termed as malware. Malware enters a user system through a network vulnerability when a user clicks a link or an email attachment that are not genuine and an unknown program installs in the system without the user being aware about the same (Rieck, Holz, Willems, Dussel, & Laskov, 2008). Once a malware enters a system, it can perform the following tasks:

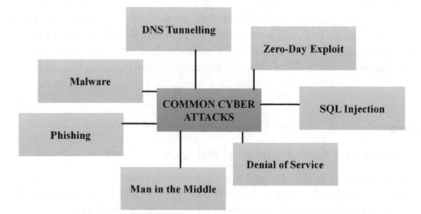

FIGURE 5.1
Common cyber-attacks.

- Block access to crucial module of the network (ransomware).
- Install unwanted apps and programs that can be used to collect information and private data of the user.
- Transmit the collected sensitive data from the hard drive to the outside world (spyware).
- Render the system dead by disrupting vital components of the system.

5.2.1.2 Phishing

Phishing is a process of sending malicious links, which seem to be genuine and valid to the user. When the user clicks on such links, they are taken to a web page that appear to be similar to some reputed source. The main aim of phishing is to steal users' data like login information and debit and credit card details or to install a backdoor to the victims system. It is a very common type of cyber-attack (Hong, 2012).

5.2.1.3 Man-in-the-middle Attack

Man-in-the-middle (MitM) attack is also called eavesdropping attack since the attacker sits in between the two communicating parties listening to all their conversations. The main aim of the attack is to gain sensitive information from the two parties and steal data (Mallik, 2019).

MitM attacks can occur on the following two scenarios:

- When using open public Wi-Fi networks, malicious users get in between a user device and the internet without the user knowing. All the information that the user passes on the network goes through the attacker.
- If a malware is activated on a computer, the attacker gets the liberty to install and load any software to steal all the victim's information.

5.2.1.4 Denial-of-service Attack

In a denial-of-service (DoS) attack, the attacker sends thousands of unwanted requests increasing the traffic on the servers and the networks. This leads to the server not being able to respond to the legitimate requests made on it. This leads to the user to wait (Moore, Shannon, Brown, Voelker, & Savage, 2006). If the attacker uses more than one system to flood the servers and networks, it is known as a distributed-denial-of-service (DDoS) attack.

5.2.1.5 Structured Query Language Injection

As the name suggests, a structured query language (SQL) injection takes place when a malicious user inserts an unwanted and harmful piece of code into the search query of a server that is operating on SQL. This pushes the server to show data which is not meant to be displayed in normal conditions (Halfond, Viegas, & Orso, 2006).

5.2.1.6 Zero-day Exploit

When the users report some issues to the developers and the developers acknowledge and announce the vulnerability and an attack takes place before a patch or solution is implemented, it is called a zero-day exploit. Attackers make the vulnerability as their target and get into the system to steal the data. This type of threat detection requires continuous monitoring and awareness (Bilge & Dumitras, 2012).

5.2.1.7 DNS Tunneling

In this type of attack, the attacker encodes a given code of data of other programs or protocol in DNS queries and responses. Mostly, data payloads are attached to the compromised DNS server and are used by a server or other applications (Van Leijenhorst, Chin, & Lowe, 2008).

5.2.2 Cyber-Attack Trends

In this section, we have listed a few industries most reluctant to cyber-attacks. Figure 5.2 displays the cyber-attack trends.

5.2.2.1 Attacks Are on the Rise on Software Supply Chain

In this type of attacks, the hacker somehow installs a malicious code into the original software by changing one of the main components of the software and renders the complete system weak (Ellison, Goodenough, Weinstock, & Carol, 2010). The physical links in a software supply chain are only as strong as their weakest link.

These types of attacks can be broadly classified in to two classes. The first class consists of attacks with one aim of compromising a carefully defined target and studying the suppliers in order to find the weakest link of the chain to enter the system for further attack. For example, SolarWinds hack enabled the attackers to exploit a large number of prominent and prestigious companies.

The other class includes the use of software supply chains to search and compromise as many weak links as possible in a wide distribution area. One of a very recent such attack was the Kaseya hack. In this attack, the cybercriminals gained access to the customer network by exploiting the software component used by managed services providers (MSPs).

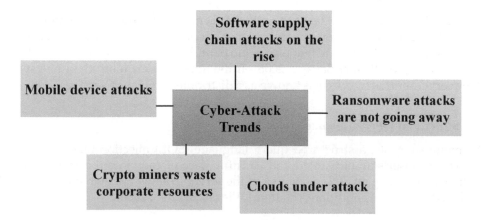

FIGURE 5.2
Cyber-attack trends.

5.2.2.2 Ransomware Attacks Are There to Stay

Ransomware has grown to become one of the most common type of cybersecurity concern for a large number of organizations. The ransomware malware makes the corporate operations, systems, and data at huge risks, and recovery of the corrupted data requires huge investments and sometimes recovery is not even possible (Gazet, 2008).

Ransomware attacks have been very successful in the recent times. This has led to growth and invention of much more innovative and harder to decipher attacks. Because these types of attacks are easy to conduct and are very profitable, they continue to grow at an unexpected rate. The ransomware attackers continue to find ways to refine the techniques and the way in which the attacks are conducted.

5.2.2.3 Clouds Systems Are Also Under Attack

As the acceptance rate of public cloud architecture increases, the number of cyber-attacks targeting these platforms has also increased. These clouds contain sensitive and personal data of the users. Vulnerabilities and poor management of these resources is still the most eminent threat that catches the eyes of the hackers. This is one of the prominent causes for a large number cyber-attacks leading to data theft throughout the world (Manoj & Bhaskari, 2016).

Cloud crypto mining operations have gained popularity because of the latest and upgraded techniques which are well capable of eluding the basic cloud security protocols (Krishan, Saketh, & Tej, 2015).

5.2.2.4 Cryptominers Waste Corporate Resources

Cryptomining or cryptojacking is a process which is designed to carry out cryptocurrency mining on the machines to which the hackers have got access, also known as "infected machines". These operations consume immense CPU resources in turn consuming large amount of electric power to carry out the mining process, in turn, making financial profits for the cyber criminals (Williamson, 2018).

The widespread interest of the masses in cryptocurrencies has given rise to the cryptomining malwares which is increasing day by day. The most recent non-fungible tokens (NFTs), provide ownership to the digital cryptocurrencies which has increased the activities of the cybercriminals to manifolds (Dowling, 2021).

5.2.2.5 Attacks on Mobile Device

The current COVID-19 pandemic saw a massive increase in the usage of mobile devices as necessity of work from home and flexible workspace came into being (Alam, Parveen, & Khan, Role of Information Technology in Covid-19 Prevention, 2020). Personal computers and smart devices became the center for all the business works and operations which in turn were seen as an opportunity by the cybercriminals (Mobile, 2021).

It was reported that more than 46% organizations have reported a cybercriminal activity involving the download of a malicious application on the devices used for corporate work and tasks. With high number of companies starting to offer remote work capabilities, the workers and employees need to be vigilant because there is a surge in the number of mobile malwares and phishing attacks that target the users via SMS and emails.

5.2.3 Cyber-Physical Attacks: A Growing Global Peril

In order to gain knowledge about CPA, we first require to know about CPS. CPS comprises interacting digital, analog, physical, and human components engineered for function through integrated physics and logic (Pasqualetti, Dorfler, & Bullo, 2011). CPS are a combination of three distinct processes: computation, networking, and physical processes. CPS are controlled and monitored with the help of embedded systems and networks with the help of feedback loops where the physical processes affect computations and vice versa. CPS is the building foundation of the evolution of Business 3.0 leading to Business 4.0 (Alam & Khan, Business 4.0-A New Revolution, 2020). It has a major role in transforming the global manufacturing process.

5.2.4 Types of Cyber-Physical Attacks

In this section, we have listed some CPA. Figure 5.3 displays some CPA.

5.2.4.1 Zero-day Attacks

When the users report some issues to the developers and the developers acknowledge and announce the vulnerability and an attack takes place before a patch or solution is implemented, it is called a zero-day exploit. Attackers make the vulnerability as their target and get into the system to steal the data. This type of threat detection requires continuous monitoring and awareness (Bilge & Dumitras, 2012).

5.2.4.2 Eavesdropping Attacks (MitM)

Man-in-the-middle (MitM) attack is also called eavesdropping attack since the attacker sits in between the two communicating parties listening to all their conversations. The main aim of the attack is to gain sensitive information from the two parties and steal data (Mallik, 2019).

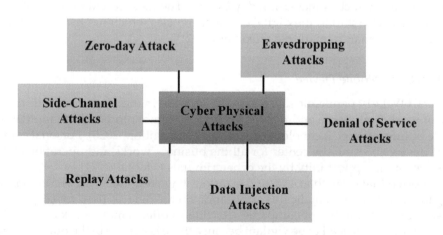

FIGURE 5.3
Cyber-physical attacks.

MitM attacks can occur on the following two scenarios:

- When using open public Wi-Fi networks, malicious users get in between a user device and the internet without the user knowing. All the information that the user passes on the network goes through the attacker.
- If a malware is activated on a computer, the attacker gets the liberty to install and load any software to steal all the victim's information.

5.2.4.3 Denial-of-Service Attacks

In a denial-of-service (DoS) attack, the attacker sends thousands of unwanted requests increasing the traffic on the servers and the networks. This leads to the server not being able to respond to the legitimate requests made on it. This leads to the user to wait (Moore, Shannon, Brown, Voelker, & Savage, 2006). If the attacker uses more than one system to flood the servers and networks, it is known as a distributed-denial-of-service (DDoS) attack.

5.2.4.4 Data Injection Attacks (SQL Injection)

As the name suggests, a structured query language (SQL) injection takes place when a malicious user inserts an unwanted and harmful piece of code into the search query of a server that is operating on SQL. This pushes the server to show data which is not meant to be displayed in normal conditions (Halfond, Viegas, & Orso, 2006).

5.2.4.5 Replay Attacks

In a replay attack, the cybercriminal eavesdrops on a safe and secure communication over a network. Then, the criminal intercepts the data received. It can then be fraudulently delayed or be resent in order to misguide the user into doing what the criminal wants (Malik, Malik, & Baumann, 2019). This attack can be very dangerous because the altered data appear to be from a recognized and trustworthy party.

5.2.4.6 Side-Channel Attacks

A side channel attack is any attack which depends on information obtained by the deployment of a computer system in place of finding weaknesses in the implementing software programme or the algorithm. It can be done by adding a simple chip with malicious code in it, by measurement and analysis of the physical parameters (Tiri, 2007).

5.2.5 Cybersecurity

Cybersecurity is the notion of protecting and securing computer systems, servers, data, databases, portable mobile devices, and networks and components from unwanted access and malicious attacks. It is also termed as information technology security or electronic information security. It can be applied in a number of environments ranging from big businesses to a single mobile of IoT smart device. A few common categories are listed below (Von Solms & Van Niekerk, 2013) (Geers, 2011). Figure 5.4 depicts some cybersecurity categories.

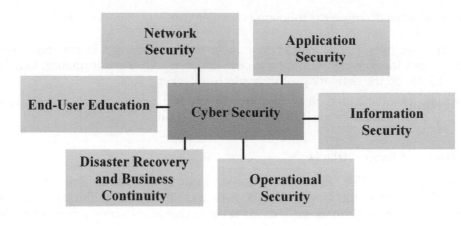

FIGURE 5.4
Cybersecurity.

5.2.5.1 Security of the Network

Securing the network means safeguarding and protecting a computer network and other networking devices from unauthorized access and malicious users.

5.2.5.2 Security of the Application

Securing the application means safekeeping and protecting the software component of the system. A malicious user can compromise an application and gain access to the data which the software needs to protect. Security paradigm should be implemented in the design phase itself, well before the application is installed.

5.2.5.3 Information Security

The work of the information security module is to protect the integrity and privacy of the data in the storage devices as well as when the data are in transit from node to node.

5.2.5.4 Operational Security

The operational security module deals with the process and decisions implemented for handling and protecting the valuable data. All the user permissions and network accessing abilities are handled in this module. How and where to store the data is also looked after in this module

5.2.5.5 Disaster Recovery and Business Continuity

This module explains how the business or user will counter to the cybersecurity event. It also defines the functionality of the business in case data loss or loss of operations occurs. It defines the policies of how the business will operate during the crisis and how it will need to recover and come back to the similar operating capacity as before the cyber-attack event.

5.2.5.6 End-user Education

This module explains the cybersecurity of the end user of the product or services. Giving user the knowledge about the threats and risks which can happen to the product and services is the foremost task. In the event when the user devices are compromised, the business needs to walk the end-user through the event and not leaving the user handle the situation in an unprofessional manner.

5.2.6 Types of Cyber-threats

The cyber-threats are three-fold:

Cybercrime: It involves a single actor or a group targeting a system for some financial gain or to cause some kind of disruption in the business.

Cyber-attack: It includes a political motivation to steal and gather information.

Cyberterrorism: The main aim of cyberterrorism is to sabotage computer and electronic systems to cause a sudden anxiety and/or terror.

Some of the well-known cyber-threats are listed below:

- Malware
- SQL injection
- Phishing
- Man-in-the-middle attack
- Denial-of-service attack

5.3 IoT

IoT is an interconnected system of wireless or wired devices like sensors, actuator, smart devices, artificial intelligence (AI) voice controllers, etc. which collect data from their surroundings and have the ability to send data to the next node over a network without the need of any human interaction. Furthermore, the data are either consumed at the very next node or is transferred to the cloud for processing and analysis (Alam, Khan, & Tanweer, IoT in Smart Cities: A survey, 9 May 2020).

The word 'thing' in internet of things can be any device which can be connected to the internet and can be assigned a unique IP address. It can include a smart wearable band, a smart overspeed detector camera, a farm animal with a transponder hanging in its neck, And sensors present in a smart-vehicle or an autonomous vehicle.

Recently, more and more industries have been adopting IoT and smart devices technology to gain more efficiency and productivity. It is helping them in better understanding of their customers and their needs and deliver better and enhanced services. These smart devices are helping them make smart choices and AI is being used for decision-making. All these are increasing the value of the business (Alam & Khan, Internet of Things as key Enabler for Efficient Business Processes, 2019).

5.3.1 How Does IoT Work?

IoT devices consist of embedded systems, including microprocessors, hardware to communicate with other nodes, sensors, etc. These devices and sensors are either data generators or data collectors. They collect the information from their surroundings. Some devices are also capable to act as per the data collected by them. These devices further connect to the IoT gateway or node or other devices on the edge, which in turn send the data to the cloud to be analyzed further. These devices are also capable to communicate to other devices present around them. They can share relevant information and act accordingly. The IoT devices require no human intervention, but humans can interact with these devices if need be. The main human interaction includes setting up the device, setting up the instructions, and reading and accessing the collected data as and when required (Mahler & Westergren, 2019). Figure 5.5 depicts the working of IoT.

The IoT application deployed, which helps in setting up these devices, have full control on the various protocols used by the devices such as communication, networking, and connectivity.

AI and machine learning (ML) are being used extensively with IoT nowadays. These technologies are making data collection and data processing easier, efficient, and dynamic (Anuj, Alam, & Khan, 2021).

5.3.2 Importance of IoT

IoT is helping the community in living and working smart and have complete command on their lives. It is providing smart devices to automate one's daily tasks and routines and keeping them safe and protected and making our lives easier. Similarly, IoT is very essential to business as well. IoT is providing business with better insights and recommendations, it is making manufacturing, logistics, marketing, construction, and other business more efficient, easier, quicker, and effective. They are helping the engineers keep an eye on the health and condition of the huge machines, engines, bridges, etc. in real time and also provide alerts if any malfunction is going to occur well in advance (Wang, Valerdi, Zhou, & Li, 2015).

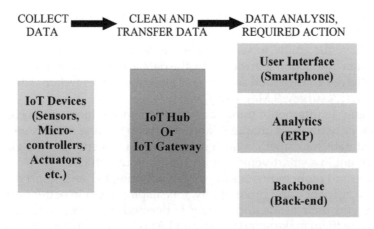

FIGURE 5.5
Working of IoT.

IoT is helping industries to automate almost all of their processes which in turn is drastically reducing the labor costs they bear. It is helping in cutting down on the waste generated and is improving the delivery of the service process. Expenses have reduced, and the customers are being offered transparency in their business and transactions.

Lately, IoT has become one of the most significant technological advancement of our everyday lives, and at current acceptance rate, it is just a matter of time as more and more business will be willing to adopt this technology and learn the true potential of the connected devices.

5.3.3 Advantages of IoT

Since the inception of IoT, organizations and industries have gained a lot. The benefits of IoT can be specific to some industries and some can be applicable to a number of organizations. (Dlamini & Johnston, 2016). Below we have listed some of the most common benefits of the technology:

- Helps in monitoring all the process of the business
- Helps in improving customer experience
- Saves money and time
- Increase productivity of the employees
- Helps in adapting and integrating new and refined business models
- Helps in making future proof and profitable decisions
- Helps in increasing revenue.

IoT is pushing business in rethinking the course of action of their business and helps in improving their strategies for the betterment of their business. Nowadays, IoT has found a home in almost every possible industry be it agriculture, healthcare, retail, manufacturing, logistics, home automation, finance, etc. IoT is even helping in making our cities smarter and safer. It is pulling everything toward digital transformation.

The use of IoT in agriculture is making the jobs of the farmers very easy. Smart sensors can be deployed around the farms which would collect the related data such as the water and manure content of the soil, temperature of the soil, humidity, rainfall and other factors, which would help and automate the farming techniques. Animals can be tracked with their tags in their neck that would prevent animals from getting lost.

IoT can also be used to monitor infrastructure and their surroundings with the help of sensors and cameras. They can monitor and report unwanted changes in the construction of buildings, bridges, and many other infrastructures. This helps in saving expenses due to over damage, saves time, and keeps the managers aware of the condition of the structure.

Smart homes are the latest trend these days. Smart devices are being used to monitor and control all the electrical and mechanical systems of the buildings. Smart cities is a great prospect in making the life of citizens comfortable and easy. It will help in reducing waste and energy consumption and make traveling and healthcare faster and cheaper (Alam, Khan, Siddique, Wiquar, & Anwar, 2020).

IoT is spreading its wings in every industry, including energy, finance, manufacturing, retail, hospitality, and transportation and logistics.

5.3.4 Pros and Cons of IoT

In this section, we have discussed some of the pros and cons of IoT (Ploennings, Cohn, & Stanford-Clark, 2018).

Listed below are some advantages of IoT:

- It provides the capability to get access to required information at any time on any device and from anywhere.
- It improves the quality of communication between various connected devices.
- It saves time and money needed in sending and receiving data over the internet.
- Automation of daily or frequent tasks drastically improves the quality of business and reduces any human intervention.

Listed below are some disadvantages of IoT:

- With increase in the number of smart devices, more and more data are generated and more and more data are shared between the devices. With increase in the information and devices, the possibility of an attack of a hacker also increases manifold.
- Huge enterprises and business may require huge number of smart devices to carry on a single task. Managing these devices and collecting the data may become challenging.
- Because of domino effect, if a single node goes down, every node can follow it and the entire system can go down.
- Currently, there is no universally accepted standard of compatibility for IoT. This makes devices from different manufacturers to communicate with one another.

5.3.5 Some IoT Standards

In this section, we discuss some the well-known IoT standards. Figure 5.6 depicts some IoT standards.

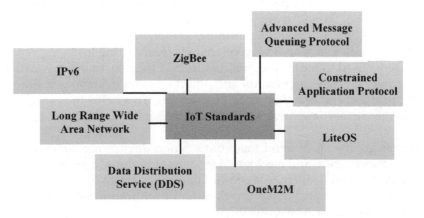

FIGURE 5.6
Some IoT standards.

5.3.5.1 IPv6

IPv6 stands for Internet Protocol Version 6. It is an internet layer protocol which uses packet-switched internetworking covering multiple IP networks. It helps a low-powered device to connect to the internet and communicate with other similar devices. It operates in 2.4 GHZ frequency range with 250 kbps transfer rate. Bluetooth low energy (BLE) and Z-Wave (for home automation) are examples (Huitema, 1995).

5.3.5.2 ZigBee

It is a low-data rate wireless network which uses low-power. It operates in 2.5 GHz frequency range at 250 kbps speed. One thousand twenty-four is the maximum number of nodes allowed with a range of up to 200 m. It uses 128-bit AES encryption. (Ramya, Shanmugaraj, & Prabakaran, 2011).

5.3.5.3 LiteOS

It is a lightweight, Unix-based operating system mainly used with sensor devices. Smartphones, wearable devices, smart home devices, and smart vehicles are all supported by LiteOS (Cao, Abdelzaher, Stankovic, & He, 2008).

5.3.5.4 OneM2M

M2M stands for machine to machine. OneM2M was built to develop technical specifications which would address the requirement of a common M2M service layer that can be embedded in small objects with a number of hardware and software (Swetina, Lu, Jacobs, Ennesser, & Song, 2014).

5.3.5.5 Data Distribution Service for Real-Time Systems (DDS)

It is the first open international IoT middleware standard for real-time, scalable, and high-performance M2M communication (Kang, Kapitanova, & Son, 2012).

5.3.5.6 Advanced Message Queuing Protocol (AMQP)

AMQP is an open standard application layer protocol for asynchronous messaging through wires. It is message oriented, uses queuing, and is reliable and secure (Appel, Sachs, & Buchmann, 2010).

5.3.5.7 Constrained Application Protocol (CoAP)

CoAP is an application layer protocol specifically designed for resource constrained IoT devices. It was proposed by IEFT (Bormann, Castellani, & Shelby, 2012).

5.3.5.8 Long Range Wide Area Network (LoRaWAN)

It is a network protocol for wide area networks (WANs), intended for wireless battery-operated devices in large networks such as smart cities, with billions of low-powered smart devices (Adelantado, et al., 2017).

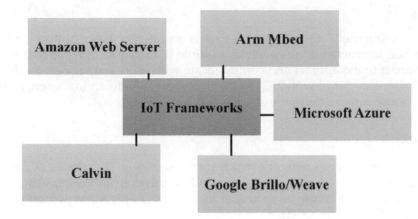

FIGURE 5.7
Some IoT frameworks.

5.3.6 IoT Frameworks

IoT frameworks have been discussed in this section. Figure 5.7 depicts some IoT frameworks.

5.3.6.1 Amazon Web Services (AWS) IoT

AWS is a cloud computing paradigm proposed by Amazon. It provides solutions and services to connect and manage billions of smart devices which are allowed to securely interact with the AWS cloud and other devices connected to the network (Bermudez, Traverso, Mellina, & Munafo, 2013).

5.3.6.2 Arm Mbed IoT

It uses Arm processors to build IoT products. It provides easily scalable framework, deeply connected, and very secure environment for all the connected smart devices. It completely integrates all the Mbed tools and services to help the users and engineers (Xiao, 2018).

5.3.6.3 Microsoft's Azure IoT Suite

It is the cloud for users as big as space stations and as small as startups and everyone in between. It means that it is highly scalable. It includes predefined set of services that enables users to interact with their smart devices, collect their data, and perform multiple operations on the data such as visualization, aggregation, and transformation and multidimensional analysis (Verma, Malla, Choudhary, & Arora, 2019).

5.3.6.4 Google's Brillo/Weave

Brillo is a light weight version of an Android OS. It provides a familiar OS with built-in support for Bluetooth and Wi-Fi and other low-power protocols used for communication. Weave is a communication layer protocol that is built into Brillo; it acts as a communication link between the cloud and the devices (Google, 2015).

5.3.6.5 Calvin

It is an IoT framework released by Ericsson. It is open source. It helps things talk to things. It includes development framework for developers and engineers, and a runtime environment for working in the running application (Personn & Angelsmark, 2015).

5.3.7 Consumer and Enterprise IoT Applications

Today, IoT is having numerous real-world uses and applications going from consumer IoT to enterprise IoT to developing, manufacturing, and industrial IoT (IIoT). Currently, IoT applications have numerous verticals, including energy, locomotive, and telecom.

In the consumer segment, smart homes are the most talked about area. Smart homes are furnished with smart devices such as smart appliances, smart thermostats, and connected heating, smart lights, and many other smart electronic devices. All of these devices can be controlled with the help of computers and smartphones from remote places. They can also be controlled via AI-based smart assistants.

The other area in the consumer segment are the wearable devices. These devices are equipped with various sensors and software which are capable of collecting the data. These devices are connected to smartphones or other such devices via Bluetooth to transfer the data for analysis. It helps in making users' life easier and more comfortable (Poongodi, Krishnamurthi, Indrakumari, Suresh, & Balusamy, 2019). Wearable devices can also be used for the safety of the public. In case of emergency, the data provided by these devices can help in locating the victim and getting to him quicker using the provided optimized routes.

In healthcare sector, IoT has numerous benefits. They help in monitoring the patients more closely and in a 24X7 manner. Doctors may not work 24 hours a day, but these devices can collect and transmit the data throughout the day and night. Smart devices are also being used as inventory management for medical equipment as well as pharmaceuticals (Mutlag, Ghani, Arunkumar, Mohammad, & Mohd, 2019).

In smart city sector, IoT is being used to construct smart buildings, smart parking systems, smart transportation, smart garbage monitoring systems, and numerus others. In smart buildings, the smart devices help in controlling and monitoring the temperature abound the premises and saving electricity when the sensors sense the room to be empty or unoccupied. (Alam, Khan, Siddique, Wiquar, & Anwar, 2020).

In the agriculture sector, IoT-based smart farming system is helping monitor the conditions of the soil such as moisture level, fertilizer level, pH level, and many others. It is also helping in making automatic irrigation system (Ruan, et al., 2019).

5.3.8 IoT Security and Privacy Issues

IoT is being accepted throughout the world in almost every sector of business. Billions of devices have already connected to the internet and this number is realized to double in less than a decade. Security and privacy of these devices are the main course of concern. As the number of devices increase, the attack surface for the attackers also increases.

In 2016, IoT witnessed its most dangerous attack up till date, Mirai. It was a botnet that penetrated into the domain name server provider Dyn. Within minutes, many websites around the globe went unreachable. It was marked as one of the biggest DDoS attack. Later, it was confirmed that the attackers gained access to the network by exploiting the less secure IoT devices on the network (Kolias, Kambourakis, Stavrou, & Voas, 2017).

Because a large number of IoT devices are connected closely, the only job a hacker needs to do is to exploit a single vulnerability to gain access to the network and manipulate or delete the data. Manufacturers should update their devices on a regular basis to protect them from such types of threats.

Hackers are not the only threat induced in IoT; user privacy is one of the next major concerns for the users. The manufacturers can use the personal user data and sell those data (Palani, 2020). Other than leaking the personal data of the user, IoT also include risk to the critical and important infrastructure, including electricity, financial services, and transportation.

5.4 Cyber-physical Attacks on IoT

CPA on IoT devices is not a new problem. As the number of IoT devices increases, they creep further and further in our daily lives. This leads to the increase in the total attack surface. Most of the sectors nowadays are getting more and more dependent on the latest smart technologies to automate their works and personal lives. Because the devices are going smart and getting connected to the internet, the physical risks because of these cyber-attacks are slowly increasing. The domino effect is most common in the IoT environment in which smart objects attack other smart objects disabling them in the process. This continues till the complete network of devices is disabled. It means if a node is attacked and the intruder has access to it, getting access to other devices on the network becomes a play of a child. When a node is compromised, all the vital details such as Wi-Fi details, location, and other such vital details are open to the intruder. In short, we can say that the security of the whole system is as strong as its weakest link.

5.4.1 Attacks on IoT Devices Have a Domino Effect

In IoT, all the sensors, actuators, and other devices produce huge amount and varieties of data on a daily basis. These data are either used, sent, or stored in vivid types of areas of an organizations IT infrastructure and data storage. Domino effect carries a negative connotation in IoT, like how a single negative step or action can lead to successive negative actions which have more impact than that or the previous one.

5.4.2 Cyber-Physical Attacks on IoT

In this section, we list a number of CPA on IoT devices. Figure 5.8 depicts some CPA on IoT.

5.4.2.1 Wireless Exploitation

As the name suggests, it needs vital information about the system, it exploits the wireless capabilities of the device, and then tries to gain remote access and control of the main router or the system. In the mean time, it disrupts the operation of the complete system (Checkoway, et al., 2011).

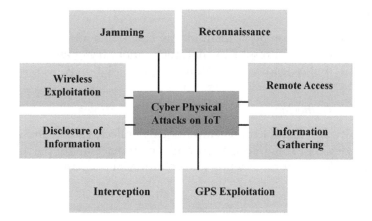

FIGURE 5.8
CPA on IoT.

5.4.2.2 Jamming

In jamming, the main aim is to block the internet access of the device or simply the access of the IoT device from the user side. In this, the state of the device changes to unavailable. This is caused by launching torrents of unwanted information or random authentication information or wireless jamming signals (Rushanan, Rubin, Kune, & Swanson, 2014).

5.4.2.3 Reconnaissance

It is also known as foot printing. In this technique, other party gathers information and foot prints about the user's computer systems and all the entities they are a part of. It is mainly done through the spread of a malware. This results in violating the data confidentiality (Miller & Rowe, 2012).

5.4.2.4 Remote Access

This is carried out by gaining remote and unauthorized access to a system. It may lead to blackout and financial losses to the user. Gaining remote access may lead to loss of sensitive and important data, which can be used to weaponize and can be used to cause cyberwarfare (Vavra & Hromada, 2015).

5.4.2.5 Disclosure of Information

Once unauthorized access is gained into a system, private as well as personal information can be gained or intercepted. This can drastically violate both privacy and confidentiality (Halperin, Heydt-Benjamin, Fu, Kohno, & Maisel, 2008).

5.4.2.6 Interception

Interception means to get in the middle of a conversation to listen to it without the two parties knowing about it. Private and personal conversations can be intercepted through compromised nodes which would lead to violation of privacy and confidentiality (Checkoway, et al., 2011).

5.4.2.7 GPS Exploitation

GPS stands for global positioning system; in this attack, the attacker gains access to the GPS device of the user, which results in continuous monitoring and tracking of the user and his locational activities leading to location privacy violation (Checkoway, et al., 2011).

5.4.2.8 Information Gathering

Hardware and software manufacturers continuously collect, gather, and analyze log files for most of their devices. This information can sometimes contain personal information which can be used illegally for marketing and commercial purposes.

5.5 Avoiding CPA on IoT

In this section, we have discussed some ways and methods to protect against the CPA which occur in our day-to-day life.

5.5.1 How to Protect Against Cyber-physical Attacks

Eliminating the above-discussed online threats and risks is an impossible task. What engineers or security officials try to do is to reduce the attacks to the minimum. Experts say that to keep a system secure and safe security testing needs to be implemented from the very beginning of the software design, that is it needs to be implemented during the entire software development life cycle (SDLC). Security needs to be a part of the system. Being a completely different entity just opens more doors for the hackers with their rapidly increasing abilities.

As the reach of internet increases and more and more devices get connected to the network, the cyber-threats and their attack surface also increases and evolves. Therefore, security systems also need to evolve to keep the users safe. Mechanisms that were sufficient in 2019 may not be efficient enough in today's world. Everything evolves, so does security. Security systems nowadays use AI and other smart technologies to keep their users and devices safe.

Cybersecurity needs to be implemented into the smart devices and technology by design. As soon as a device connects to the internet, it becomes a potential hacking opportunity. Smart devices are more than capable of taking their own decisions with the help of algorithms and AI-based knowledge; if a single point of inference is manipulated or changed, it can have drastic effects on the outcome and what the device does in the real world. Most of the above-mentioned advancements are taking place, but it is not keeping pace with the speed of IoT convergence.

Governments need to become more and more involved in implementing these security norms. Their prime area of focus lies in privacy and personally identifiable information.

5.5.2 A Cyber-attack is Preventable

Even after the manifold increase in the cyber-attacks, more than 99% of the businesses and enterprises are not protected efficiently. It should be understood that more than 90% of the cyber-attacks are preventable. The main key to cyber defense is a point-to-point and end-to-end cybersecurity framework. It needs to be a multilayered framework which includes all the devices on the networks, be it servers, end points, mobile devices, clouds, etc. A correct framework will help the management in consolidating a multiple layer security system and control policy in a single frame.

Below we have mentioned a few key actions which need to be taken in view of preventing cyber-attacks:

- Updating security regularly
- Prevention is better than detection
- Secure total network and all devices
- Implementing and using latest technology
- Threat intelligence should have the latest definitions

5.5.3 Protection for the End User

One of the most crucial part of cybersecurity is the end-user protection. Also known as endpoint protection is the attempt to stop unauthorized and malicious users or groups from attacking and compromising the end-user devices such as desktops, laptops, smartphones, and IoT devices. Often, the less technical user is the one who unknowingly downloads and installs a malicious piece of code into their devices which becomes the source of cyber threat on the network. Cryptography is the main technique of data protection for cybersecurity. It encrypts the sensitive data and protects it at all times, be it in transit or on the hard drive.

Security software installed by end users on their systems scans and protects their computers from dangerous programs and malicious codes. Whenever such a code is detected or found, the program is either put under quarantine or is completely deleted from the system. They are even equipped to scan the Master Boot Record (MBR) where the malicious codes can hide which if activated can wipe all the data on the system.

The main aim of the electronic security protocol is real-time detection of suspicious codes and malwares. They make use of heuristics and make analysis of the behavior of the suspected program and its code to protect users from worms, adware, trojans, and viruses. They capture the suspected programs into secure bubbles away from the main system or networks to study the behavior of the malicious code for future protection and security.

New threats and attacks are being detected on a daily basis. Security programs need to evolve in order to protect the users. New threats need new ways to counter them. End users need to know how to use these software. Tasks such as updating need to be performed regularly by the users. Users need to know that running and updating the security software are the most important aspect of any cybersecurity paradigm.

5.5.4 Cyber Safety Tips—Protect Oneself Against Cyber-attacks

In this section, we have listed a few cybersecurity tips. Figure 5.9 depicts some cybersecurity tips.

5.5.4.1 Update Software and Operating System Regularly

This includes updating the software component of the devices to the latest available official firmware. It may or may not include security and vulnerability patches. Updating means the end-user benefits from the latest security patches.

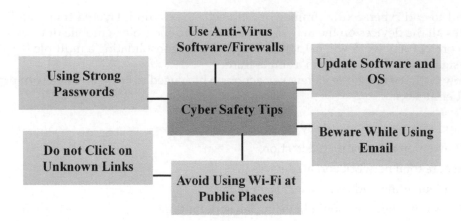

FIGURE 5.9
Cyber safety tips.

5.5.4.2 Use Anti-virus Suite

Total security suits such as Bit Defender Total Security, Kaspersky Total Security, and Norton Total Security once installed on a user device, constantly monitor and protect the device from malicious programs and threats. They detect and remove threats from the device. The user should keep the anti-virus software updated with latest definitions for the top level of protection.

5.5.4.3 Use Strong Passwords

Every device, nowadays, has the option to be locked and unlocked only with the user-provided passphrase or code. Users should use strong passwords that should not be easily guessable to have the maximum protection.

5.5.4.4 Email Attachments from Unknown Senders Should Not Be Entertained

Email attachments should not be opened until and unless you trust the sender. Opening attachments that are infected with malware may corrupt the files and data on the user device. These could even end with locking the user out of their own devices.

5.5.4.5 Links in Emails from Unknown Senders or Unfamiliar Websites Should Not Be Clicked

Emails from unknown senders, sometimes referred to as spams, should not be opened. This is the most common way to spread a malware. Clicking on such links may lead to websites that are not genuine and are made to carry on the phishing attacks.

5.5.4.6 Unsecure Wi-Fi Networks in Public Places Should Be Avoided

Connecting to unsecure and open networks leaves the user data vulnerable and it can be compromised easily. These data can be stolen with mechanisms like MitM attacks.

5.6 Conclusion

In this chapter, we have tried to explain about CPS and CPA. Furthermore, we discussed about IoT and the severe effects of CPA on the IoT sensors, actuators, and other devices.

Manufacturers require to make smart devices to implement and utilize 'reasonable' measures to make sure that the devices and systems they manufacture have adequate security. Physical damages from insecure and vulnerable devices could open a large number of avenues of legal liability, ranging from civil suits brought by injured users, business partners and shareholders to more aggressive sanctions by government to criminal liability.

To avoid the legal punishments, manufacturers need to take measures that should be fundamental. But such measures are a long way from universal. In addition, having inbuilt security measures into the products before they hit the market would be a much easier and robust measure. Manufacturers need to be transparent in contractual relationships as to what risks are being borne by the end users. Furthermore, what are the obligations of each business partner, software provider, hardware manufacturer, network component, and in some cases, what is a user responsibility needs to be justified in advance.

To limit user liability, companies need to show that they take security more seriously than the users of their products do. That is still very obviously a long way from reality which needs to be worked upon in the future.

References

Adelantado, F., Vilajosana, X., Tuset-Peiro, P., Martinez, B., Melia-Segui, J., & Watteyne, T. (2017). Understanding the Limits of LoRaWAN. *IEEE Communications Magazine*, *55*(9), 30–34. doi:10.1109/MCOM.2017.1600613

Alam, M., & Khan, I. R. (2019). Internet of Things as Key Enabler for Efficient Business Processes. *SSRN Electronic Journal*. doi:10.2139/ssrn.3806408

Alam, M., & Khan, I. R. (2020). Business 4.0-A New Revolution. In *Information Technology for Management*. Delhi: KD Publications. doi:10.6084/m9.figshare.14369636

Alam, M., Khan, I. R., & Tanweer, S. (2020, 9 May). IOT in Smart Cities: A Survey. *Juni Khyat*, 89–101. doi:10.6084/m9.figshare.14329718

Alam, M., Khan, I. R., Siddique, S. A., Wiquar, R., & Anwar, H. (2020). IoT and AI as key enabler of growth of smart cities. *2nd International Conference on ICT for Digital, Smart and Sustainable Development*. New Delhi. doi:10.4108/eai.27-2-2020.2303467

Alam, M., Parveen, R., & Khan, I. R. (2020). Role of Information Technology in Covid-19 Prevention. *International Journal of Business Education and Management Studies*, 65–75. doi:10.6084/m9.figshare.14369627.v1

Amour, L. S. (2012). The Internet: An Unprecedented and Unparalleled Platform for Innovation and Change. In L. S. Amour (ed.), *The Global Innovation Index* (pp. 157–162). WIPO.

Anderson, R., Barton, C., Bohme, R., Clayton, R., Van Eeten, M. J., Levi, M., … Savage, S. (2013). Measuring the Cost of Cybercrime. In R. Bohme (ed.), *The Economics of Information Security and Privacy* (pp. 265–300). Berlin: Springer, Berlin, Heidelberg. doi:10.1007/978-3-642-39498-0_12

Anuj, R., Alam, M., & Khan, I. R. (2021). Smart garbage monitoring system using IoT. *2nd International Conference on Emerging Trends in Mathematical Sciences & Computing (IEMSC-21)*. Kolkata: West Bengal. doi:10.2139/ssrn.3902056

Appel, S., Sachs, K., & Buchmann, A. (2010). Towards benchmarking of AMQP. *Proceedings of the Fourth ACM International Conference on Distributed Event-Based Systems*. New York: ACM Press. doi:10.1145/1827418.1827438

Bermudez, I., Traverso, S., Mellina, M., & Munafo, M. (2013). Exploring the cloud from passive measurements: The Amazon AWS case. *Proceedings IEEE INFOCOM* (pp. 230–234). Turin, Italy: IEEE. doi:10.1109/INFCOM.2013.6566769

Bilge, L., & Dumitras, T. (2012). Before we knew it: An empirical study of zero-day attacks in the real world. *Proceedings of the 2012 ACM conference on Computer and Communications Security (CCS '12)* (pp. 833–844). New York: ACM. doi:10.1145/2382196.2382284

Bormann, C., Castellani, A. P., & Shelby, Z. (2012). CoAP: An Application Protocol for Billions of Tiny Internet Nodes. *IEEE Internet Computing, 16*(2), 62–67. doi:10.1109/MIC.2012.29

Cao, Q., Abdelzaher, T., Stankovic, J., & He, T. (2008). The LiteOS operating system: towards unix-like abstractions for wireless sensor networks. *International Conference on Information Processing in Sensor Networks (IPSN 2008)* (pp. 233–244). St. Louis, MO, USA: IEEE. doi:10.1109/IPSN.2008.54

Checkoway, S., McCoy, D., Anderson, D., Kantor, B., Shacham, H., Savage, S., … Kohno, T. (2011). Comprehensive experimental analyses of automotive attack surfaces. *Proceedings of the USENIX Security Symposium* (pp. 77–92). San Francisco.

CISCO. (2019, October). *CISCO Cybersecurity Series 2019*. Retrieved from CISCO: https://www.cisco.com/c/m/en_au/products/security/offers/cybersecurity-reports.html

Denning, P. J., & Denning, D. E. (2010). Discussing Cyber Attack. *Communications of the ACM, 53*(9), 29–31. doi:10.1145/1810891.1810904

Dlamini, N. N., & Johnston, K. (2016). The use, benefits and challenges of using the Internet of Things (IoT) in retail businesses: A literature review. *International Conference on Advances in Computing and Communication Engineering (ICACCE)* (pp. 430–436). Durban: IEEE. doi:10.1109/ICACCE.2016.8073787

Dowling, M. (2021). Fertile LAND: Pricing Non-Fungible Tokens. *Finance Research Letters*, 102096. doi:10.1016/j.frl.2021.102096

Ellison, R. J., Goodenough, J. B., Weinstock, C. B., & Carol, W. (2010). *Evaluating and Mitigating Software Supply Chain Security Risks*. Pittsburgh: Carnegie-Mellon Univ Pittsburgh Pa Software Engineering Inst.

Elsisi, M., Tran, M.-W., Mahmoud, K., Mansour, D.-E. A., Lehtonen, M., & Darwish, M. M. (2021). Towards Secured Online Monitoring for Digitalized GIS Against Cyber-Attacks Based on IoT and Machine Learning. *IEEE Access, 9*, 78415–78427. doi:10.1109/ACCESS.2021.3083499

Gazet, A. (2008). Comparative Analysis of Various Ransomware Virii. *Journal in Computer Virology, 6*, 77–90. doi:10.1007/s11416-008-0092-2

Geers, K. (2011). *Strategic Cyber Security*. Estonia: NATO Cooperative Cyber Defence Centre of Excellence.

Google. (2015). *Google's Internet of Things solution*. Retrieved from Google Developer: https://developers.google.com/iot

Halfond, W. G., Viegas, J., & Orso, A. (2006). A classification of SQL-injection attacks and countermeasures. *Proceedings of the IEEE International Symposium on Secure Software Engineering* (pp. 13–15). IEEE.

Halperin, D., Heydt-Benjamin, T. S., Fu, K., Kohno, T., & Maisel, W. H. (2008). Security and Privacy for Implantable Medical Devices. *IEEE Pervasive Computing, 9*(1), 30–39. doi:10.1109/MPRV.2008.16

Hampton, K. N. (2010). Internet Use and the Concentration of Disadvantage: Glocalization and the Urban Underclass. *American Behavioral Scientist, 53*(8), 1111–1132. doi:10.1177/0002764209356244

Hendrik, S., & Mochalski, K. (2009). *Internet Study 2008/2009*. Leipzig, Germany: Ipoque Report 37.

Hong, J. (2012). The State of Phishing Attacks. *Communications of the ACM, 55*(1), 74–81. doi:10.1145/2063176.2063197

Huitema, C. (1995). *IPv6: The New Internet Protocol*. United States: Prentice-Hall, Inc.

Jahromi, A. N., Karimipour, H., Dehghantanha, A., & Choo, K.-K. R. (2021). Toward Detection and Attribution of Cyber-Attacks in IoT-Enabled Cyber–Physical Systems. *IEEE Internet of Things Journal, 8*, 13712–13722. doi:10.1109/JIOT.2021.3067667

Kang, W., Kapitanova, K., & Son, S. H. (2012). RDDS: A Real-Time Data Distribution Service for Cyber-Physical Systems. *IEEE Transactions on Industrial Informatics, 8*(2), 393–405. doi:10.1109/TII.2012.2183878

Kolias, C., Kambourakis, G., Stavrou, A., & Voas, J. (2017). DDoS in the IoT: Mirai and Other Botnets. *Computer, 50*(7), 80–84. doi:10.1109/MC.2017.201

Krishan, H., Saketh, S., & Tej, V. (2015). Cryptocurrency Mining-Transition to Cloud. *International Journal of Advanced Computer Science and Applications, 6*(9), 115–124. doi:10.14569/IJACSA.2015.060915

Li, F., Shi, Y., Shinde, A., Ye, J., & Song, W. (2019). Enhanced Cyber-Physical Security in Internet of Things Through Energy Auditing. *IEEE Internet of Things Journal, 6*(3), 5224–5231. doi:10.1109/JIOT.2019.2899492

Mahler, V., & Westergren, U. H. (2019). Working with IoT – A Case Study Detailing Workplace Digitalization Through IoT System Adoption. In L. Strous, & V. Cerf (eds.), *IFIP Advances in Information and Communication Technology* (pp. 178–193). Cham: Springer. doi:10.1007/978-3-030-15651-0_15

Malik, K. M., Malik, H., & Baumann, R. (2019). Towards vulnerability analysis of voice-driven interfaces and countermeasures for replay attacks. *IEEE Conference on Multimedia Information Processing and Retrieval (MIPR)* (pp. 523–528). San Jose, CA: IEEE. doi:10.1109/mipr.2019.00106

Mallik, A. (2019). Man-in-the-Middle-Attack: Understanding in Simple Words. *Cyberspace: Jurnal Pendidikan Teknologi Informasi, 2*(2), 109. doi:10.22373/cj.v2i2.3453

Manoj, S. A., & Bhaskari, L. (2016). Cloud Forensics-A Framework for Investigating Cyber Attacks in Cloud Environment. *Procedia Computer Science, 85*, 149–154. doi:10.1016/j.procs.2016.05.202

Miller, B., & Rowe, D. (2012). A survey SCADA of and critical infrastructure incidents. *Proceedings of the 1st Annual Conference on Research in Information Technology – RIIT '12* (pp. 51–56). New York: ACM. doi:10.1145/2380790.2380805

Mobile, T. (2021, Feb 24). *5 Reasons Hackers Target Mobile Devices and How to Stop Them*. Retrieved from Forbes: https://www.forbes.com/sites/tmobile/2021/02/24/5-reasons-hackers-target-mobile-devices-and-how-to-stop-them/?sh=7f0af91a7b28

Moore, D., Shannon, C., Brown, D. J., Voelker, G. M., & Savage, S. (2006). Inferring Internet Denial-of-Service Activity. *ACM Transactions on Computer Systems, 24*(2), 115–139. doi:10.1145/1132026.1132027

Mutlag, A. A., Ghani, M. K., Arunkumar, N., Mohammad, M. A., & Mohd, O. (2019). Enabling Technologies for Fog Computing in Healthcare IoT Systems. *Future Generation Computer Systems, 90*, 62–78. doi:10.1016/j.future.2018.07.049

Palani, N. (2020). ONE-GUI Designing for Medical Devices & IoT Introduction. In P. S. Shanmugam, L. Chokkalingam, & P. Bakthavachalam (eds.), *Trends in Development of Medical Devices* (pp. 17–34). Academic Press. doi:10.1016/B978-0-12-820960-8.00002-2

Pasqualetti, F., Dorfler, F., & Bullo, F. (2011). Cyber-physical attacks in power networks: Models, fundamental limitations and monitor design. *50th IEEE Conference on Decision and Control and European Control Conference*, (pp. 2195–2201). Orlando, FL. doi:10.1109/CDC.2011.6160641

Personn, P., & Angelsmark, O. (2015). Calvin – Merging Cloud and IoT. *Procedia Computer Science, 52*, 210–217. doi:10.1016/j.procs.2015.05.059

Ploennings, J., Cohn, J., & Stanford-Clark, A. (2018). The Future of IoT. *IEEE Internet of Things Magazine, 1*(1), 28–33. doi:10.1109/IOTM.2018.1700021

Poongodi, T., Krishnamurthi, R., Indrakumari, R., Suresh, P., & Balusamy, B. (2019). Wearable Devices and IoT. In V. Balas, V. Solanki, R. Kumar, & M. Ahad (eds.), *A Handbook of Internet of Things in Biomedical and Cyber Physical System* (pp. 245–273). Springer. doi:10.1007/978-3-030-23983-1_10

Ramya, C. M., Shanmugaraj, M., & Prabakaran, R. (2011). Study on ZigBee technology. *3rd International Conference on Electronics Computer Technology* (pp. 287–301). Kanyakumari, India: IEEE. doi:10.1109/ICECTECH.2011.5942102

Rana, M. M., Khan, M. R., & Abdelhadi, A. (2020). IoT Architecture for Cyber-Physical System State Estimation Using Unscented Kalman Filter. *Second International Conference on Inventive Research in Computing Applications (ICIRCA)* (pp. 910–913). Coimbatore, India: IEEE. doi:10.1109/ICIRCA48905.2020.9183350

Rieck, K., Holz, T., Willems, C., Dussel, P., & Laskov, P. (2008). Learning and Classification of Malware Behavior. In D. Zamboni (ed.), *Detection of Intrusions and Malware, and Vulnerability Assessment* (pp. 108–125). Berlin: Springer Berlin Heidelberg. doi:10.1007/978-3-540-70542-0_6

Ruan, J., Jiang, H., Zhu, C., Hu, X., Shi, Y., Liu, T., … Chan, F. T. (2019). Agriculture IoT: Emerging Trends, Cooperation Networks, and Outlook. *IEEE Wireless Communications, 26*(6), 56–63. doi:10.1109/MWC.001.1900096

Rushanan, M., Rubin, A. D., Kune, D. F., & Swanson, C. M. (2014). SoK: Security and privacy in implantable medical devices and body area networks. *IEEE Symposium on Security and Privacy* (pp. 524–539). Berkeley, CA, USA: IEEE. doi:10.1109/SP.2014.40

Sen, S., & Jayawardena, C. (2019). Analysis of cyber-attack in big data IoT and cyber-physical systems – A technical approach to cybersecurity modeling. *5th International Conference for Convergence in Technology (I2CT)* (pp. 1–7). Bombay, India: IEEE. doi:10.1109/I2CT45611.2019.9033821

Stefano Baldi, E. G. (2003). *Hacktivism, cyber-terrorism and cyberwar*. Switzerland: DiploFoundation.

Swetina, J., Lu, G., Jacobs, P., Ennesser, F., & Song, J. (2014). Toward a standardized common M2M service layer platform: Introduction to oneM2M. *IEEE Wireless Communications, 21*(3), 20–26. doi:10.1109/MWC.2014.6845045

Tiri, K. (2007). Side-channel attack pitfalls. *4th ACM/IEEE Design Automation Conference* (pp. 15–20). San Diego, CA: IEEE.

Van Deursen, A. J., & Heisper, E. J. (2017). Collateral Benefits of Internet Use: Explaining the Diverse Outcomes of Engaging with the Internet. *New Media & Society, 20*(7), 2333–2351. doi:10.1177/1461444817715282

Van Leijenhorst, T., Chin, K.-W., & Lowe, D. (2008). On the viability and performance of DNS tunneling. *The 5th International Conference on Information Technology and Applications (ICITA 2008)*. Cairns, Australia.

Vavra, J., & Hromada, M. (2015). An evaluation of cyber threats to industrial control systems. *International Conference on Military Technologies (ICMT)* (pp. 1–5). Brno, Czech Republic: IEEE. doi:10.1109/MILTECHS.2015.7153700

Verma, A., Malla, D., Choudhary, A. K., & Arora, V. (2019). A detailed study of Azure platform & its cognitive services. *International Conference on Machine Learning, Big Data, Cloud and Parallel Computing (COMITCon)* (pp. 129–134). Faridabad, India: IEEE. doi:10.1109/COMITCon.2019.8862178

Von Solms, R., & Van Niekerk, J. (2013). From Information Security to Cyber Security. *Computers & Security, 38*, 97–102. doi:10.1016/j.cose.2013.04.004

Wang, P., Valerdi, R., Zhou, S., & Li, L. (2015). Introduction: Advances in IoT Research and Applications. *Information Systems Frontiers, 17*(2), 239–241. doi:10.1007/s10796-015-9549-2

Williamson, S. (2018). Is bitcoin a Waste of Resources? *SSRN Electronic Journal*, 107–115. doi:10.20955/r.2018.107-15

Xiao, P. (2018). *Designing Embedded Systems and the Internet of Things (IoT) with the ARM Mbed*. Wiley.

You, I., Yim, K., Sharma, V., Choudhary, G., Chen, I.-R., & Cho, J.-H. (2018). On IoT misbehavior detection in cyber physical systems. *23rd Pacific Rim International Symposium on Dependable Computing (PRDC)* (pp. 89–90). Taipei, Taiwan: IEEE. doi:10.1109/PRDC.2018.00033

6

Chaos-Based Advanced Encryption Algorithm Using Affine Transformation in S-Box and Its Implementation in FPGA

V. Nandan and R. Gowri Shankar Rao

Vel Tech Rangarajan Dr Sagunthala R & D Institute of Science and Technology, Tamil Nadu, India

CONTENTS

6.1 Overview

In the data communication system, security must be concentrated. Security and the complexity of algorithms dealing with security increase because of the randomization in secret keys. Hence, recently, cryptography algorithms must be capable of balancing enormous memory and execution time based on the hardware platform. FPGA (field-programmable gate arrays), a reprogrammable device, are attractively used to implement hardware employed in encryption algorithms. AES, an efficient and cost-effective symmetric cryptographic algorithm, is broadly used in applications like ATMs, smart cards, mobiles, and internet servers to maintain data confidentiality. A secure, robust cryptosystem depends on AES substitution box, and chaotic components are proposed in this chapter. An effective pseudo chaotic number generator (PCNG), global diffusion, and block cipher are comprised. The finite field is used to define the PCNG, which eliminates the danger of depreciated security as a subsequent dynamic degradation. At the same time, numerical implementation of chaotic maps is always stated as real numbers. In the modified Bernoulli process, diffusion properties are increased efficiently by horizontal addition diffusion (HAD) and vertical addition diffusion (VAD). The strength of cryptography is tested in detail in S-Box proposed through the various benchmark standards. The results of experiments and examination of performance showed that this Chaos-based advanced encryption algorithm proposed using S-box Bernoulli process (CA-S-box-BP) achieves 96% efficiency, 64% non-linear criterion, and 41% strict avalanche effect with 98 Kbps in 75 ms.

6.2 Introduction

NIST, stands for National Institute of Standards and Technology, reviewed and published substitution cum permutation block cipher called Advanced Encryption Standard (AES). Joan Daemen and Vincent Rijmen designed this AES standard. FIPS which stands for Federal Information Processing Standards approved and announced this standard (Hosseinkhani et al., 2012). Modular arithmetic is used to generate an eight-bit S-Box, which is motivated to present non-linearity in AES encryption. Decryption is carried out by building forward S-Box and inverse S-Box (Paul et al., 2012). The specific polynomial {11B} and an additive number (63) under GF(2), which is irreducible due to modulus in GF(28), are used to generate AES standard S-Box. Rijndael uses this specific modulus and the additive constant in the original design. To generate a more dynamic S-Box (Özkaynak, 2018), it also uses other moduli and constants (Jingmei et al., 2007). The cryptographic algorithm's security is characterized by recommending some criteria of NIST. The lengthy binary sequence in the 106th order randomness is verified by 15 statistical package tests present in NIST Test Suite. This lengthy binary sequence is dedicated to the sequence's randomness in several steps; step1 is to select the generator appropriate to cryptography applications. Cryptanalysis is not substituted by statistical testing, as per the declaration of NIST (Rahim et al., 2019). Several S-Boxes are used to generate AES ciphertexts, and later, randomness and security of the S-Boxes are tested based on the specific S-Box selected (Zhun and Sun, 2018). Confusion and diffusion are the two elementary properties provided by the block encryption algorithm. The substitution box (s-box) structure, also called cryptographic components, is provided by the property of confusion in several block encryption algorithms. So, the structure of the s-box provides strength to block encryption algorithms. The s-box is designed by algebraic, pseudorandom, and heuristic methods (Zhang et al., 2018). Depending on the strong algebraic relations, s-box design techniques are often used by recent block encryption algorithms, and many new algorithms were presented by Nyberg (Ahmed et al., 2019). AES's s-box block encryption algorithm uses this method (Murugan et al., 2020). The algorithms designed by algebraic s-box are alternated by the chaos-based s-box techniques presented last decade. The similarities between cryptography and Chaos are the basis for designing s-box in chaotic systems (Guo et al., 2018). Randomness source is used in the design of chaotic systems. The chaotic system output creates S-box structures (Wang et al., 2018). The properties of cryptography AES s-box structure were worse than AES s-box structure while examining performance criteria in chaos-based s-box structures (Ping et al., 2018). For example, 112 is the AES s-box value of non-linearity (Zhu et al., 2018). Well-known features of cryptography are used in designing AES s-box, and the upper limit value was reached. The value of non-linearity in the maximum limit during the Chaos-based s-box is 106.75. Structure of the AES s-box, differential attack's resistance is measured by the Input/Output XOR (differential analysis)'s maximum value. In this structure, AES s-box, this maximum value is 4. Maximum value obtained as 4 is small and maintained as small as possible. The motivation of this chapter is to design a secure, robust cryptosystem that depends on AES substitution box and chaotic components and to test the strength of cryptography in detail in S-Box proposed through the various benchmark standards. This chapter is organized as: In Section 6.1, the background of advanced encryption and its usage in the s-box is given. In Section 6.2, the existing works related to various s-box is discussed. In Section 6.3, a new s-box is constructed based on the chaos method by including the Bernoulli concept. Section 6.4 presents the experimental analysis by applying the proposed s-box with various FPGA platforms. Finally, the chapter ends with Section 6.5, conclusion and future work.

6.3 Related Works

Many researchers developed the design of the S-box to accomplish increased non-linearity and low DP values. Robust similarities between Chaos and cryptography should be present to design the S-box. Initial conditions depending on the sensitivity, ergodicity, and mixing are some of the intrinsic properties of Chaos used by various researchers (Kocarev, 2001). 1D, 2D, and 3D chaotic maps are used to design S-boxes (Belazi et al., 2018; Tang et al., 2005). The 6Dchaotic map and artificial bee colony algorithm (Hussain et al., 2013) presented the design of the S-box. The most appropriate initial condition and control parameter values consistent with four chaotic algorithms are determined by seven various optimization algorithms (Tanyildizi and Ozkaynak, 2019).

Optimized chaotic maps designed another S-box with very low robustness than the AES s-box. Based on the perspective of cryptography, 3D plasma properties are used to design the S-box generation. The combination of chaotic map, Baker's map, along the sinusoidal chaotic map in Linear Congruence Generator (LCG) was used for the S-box design in wireless sensor networks (Yi et al., 2019). A new S-box was generated by linearly transforming through the chaotic logistic tent maps and gold sequence results while designing a cryptographic approach for the substitution-permutation network (Khan et al., 2019). By using the map of tent-logistic chaotic, a new S-box was designed. This tent-logistic chaotic map includes a new linear mapping scheme for generating the s-box (Lu et al., 2019). Image encryption is carried out by a chaotic Jaya optimization algorithm (Hayat et al., 2018) in which the properties of cryptography are accomplished better by the S-box. Some researchers introduced a few approaches to chaos results in strong s-box generation. The combination of Chaos and algebra obtains more appropriate cryptography results. A discrete–space chaotic map in a single dimension is presented (Lambić, 2020). Analysis of its dynamic properties with the existence of chaotic behavior is performed. As a result, S-boxes adopted with the best cryptographic chaotic map is generated. Secure and efficient chaotic S-Box dependent on image encryption algorithm was presented (Lu et al., 2020). The substitution key sequences and permutation are linked with plaintext image content, and this approach allows the cryptosystem for resisting a chosen-plaintext attack (CPA).

6.4 System Model

The computation of the effective AES S-box and its preliminary works are explained in this section. Figure 6.1 shows input parameters with encryption process continued by s-box construction with the assistance of the Bernoulli process and block ciphertext.

6.4.1 Encryption Process

Figure 6.2 represents the encryption process. The secret key controls PCNG, which stands for Pseudo Chaotic Number Generator, and is a sample of PCNG to process encryption, including with the operation of global diffusion and block cipher. The properties of diffusion are enhanced by global diffusion. In step 1, the operation of global diffusion with the diffusion layer is used to process the entire input plain image. HAD, which stands for Horizontal Addition Diffusion, is continued by VAD, which stands for Vertical Addition Diffusion (VAD), along with permutation layer, which depends on the Bernoulli process, is

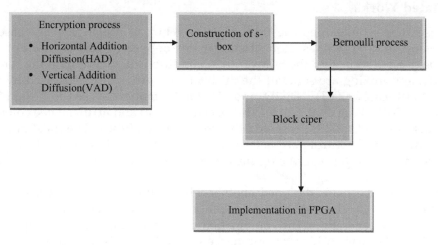

FIGURE 6.1
System architecture for proposed CA-S-box-BP.

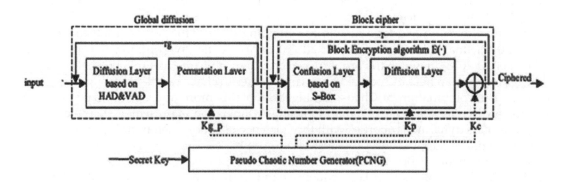

FIGURE 6.2
Encryption process of chaos-based method.

used in this global diffusion operation. PCNG provides the parameter needed for this process. The rg times this process of diffusion will be repeated. Then, global diffusion's output was divided into blocks containing 16 bytes for a block of 4 × 4bytes. For block encryption, a block cypher kernel algorithm called E(.) is used, which consists of a confusion layer based on Advanced Encryption Standard SBox, a key addition operation (XOR), and a diffusion layer. The two-dimensional cat map modified, PCNG fed XOR operator requires Kp; Kc. The final ciphered text is obtained by repeating the block cypher operation r times.

6.4.2 Construction of S-Box

The substitution box non-linear component is present in the symmetric block algorithms. An S-box negotiates the differential attack and linear attack. GF (2)8 to GF(2)8's non-linear mapping is used to generate 8 × 8 S-Box mathematically, where GF(2)8 is the vector space with 28 members, and it is signified as 8-tuples of bits from GF(2).

The following steps construct S-box.

- Assign n as16 and consider p as a prime number with the value of 101.
- Preparation of (1 × 256) S array having the elements $si = (i \times p)$ mod $(n2)$ where $1 \leq i \leq 256$.
- Run chaotic Sine map $xn + 1 = \beta$ sin $\pi \times n$ L iteratively as 1000 times for avoiding the transient effect having β = 2.1676049460705697 and initial state $x0$ = 0.84584390650801455 and chaotic Sine map in a (1 × 256) array A is stored.
- All the values are sorted in ascending order, and the latest (1 × 256) array B is stored.
- Its position value replaces every bi member in array B in bi present in A array.
- Preparation of array D (1 × 256) where di ¼ sbi.
- Rearrangement of (1 × 256) array D in $(n \times n)$ matrix D is the S- box proposed in the last step. The hexadecimal form S-Box is obtained.

S-boxes are robust against linear and differential attacks when there is high non-linearity (NSB) and low differential uniformity (ΔSB). Intended for S-Box, which is balanced: GF(2n) → GF(2n); n is even, the NSB upper bound is 2n–1–2n2–1–2 and produces SB: GF(2n)→GF(2m), and ΔSB lower bound is 2n – m + 1. The calculation for the S-Box robustness in contradiction of differential cryptanalysis is mentioned in Equation (6.1).

$$\beta = \left(1 - \frac{R}{2n}\right)\left(1 - \frac{L}{2n}\right) \tag{6.1}$$

where DDT larger value is L and first entry under the first row is excluded, and R is the nonzero count under DDT's first column with similar exclusion stated above. Freedom of Output bit is a crucial condition for S-Box; by changing the kth input bit, both output bits in the position of I and j must arise autonomously. Calculating $\rho ij(k)$; $1 \leq k \leq n$, which are the coefficients of correlation and the criteria of bit independence criterion, is provided in Equation (6.2).

$$\max[1 \leq k \leq n]pij(k) \tag{6.2}$$

6.4.3 Bernoulli Process

Let $(Tk)k \in N0$ be a multiplicative random walk defined in Equation (6.3),

$$T0: = 1, Tk = \prod_{I=1}^{K} Wik \in N \tag{6.3}$$

where $(Wk)k \in N$ are independent copies of a random variable taking values in (0,1). The random variables $(Uk)k \in N$ are independent in uniform [0,1] law, which means they are not dependent on the random walk, which is multiplicative. The Bernoulli riddles a scheme of random occupancy in which the Uk's balls are assigned above numerous boxes $(Tk, Tk - 1), k \in N$.

$$Pk: = Tk - 1 - Tk = W1W2 \cdots Wk - 1(1 - Wk) \tag{6.4}$$

As mentioned in Equation (6.4), the traditional scheme of infinite occupancy is Bernoulli sieve [8,17] having random frequencies $(Pk)k \in N$, where balls (abstract) are assigned above a box (abstract) with infinite array 1,2; for the box j hitting, Pk is provided with probability Pjo. Otherwise, the Bernoulli sieve is believed as a leader election procedure's randomized variant, appearing when the W law degenerates in a few $x \in (0, 1)$.

Based on the W law behavior, closer to 0 and 1 endpoints, empty boxes will display various asymptotics' fairly extended range.

Case $\mu < \infty$ and $\nu < \infty$: Ln come together in distribution and mean with few L having proper and non-degenerate law.

Case $\mu = \infty$ and $\nu < \infty$: Ln converges to zero in probability.

Case $\mu < \infty$ and $\nu = \infty$: There are weak convergence probable modes of Ln, appropriately centered and normalized.

Case $\mu = \infty$ and $\nu = \infty$: The asymptotics of Ln is regulated through performance ratio of the $P\{W \le x\}/P\{1 - W \le x\}$ as $x \downarrow 0$. At once, W law allocates huge mass to the neighborhood of 1, neglecting 0 is equivalent to the ratio going toward 0, and Ln turns out to be large as per the asymptotical method. Ln is resulted by weak convergence for this situation.

6.4.4 Block Cipher

Completing global diffusion, divide the diffused image to 4×4 bytes for a block. An algorithm of block encryption $E(\cdot)$ is used to process every block. To obtain a better security performance, r times the ciphering process is repeated. The AES s-box is blocked encryption algorithm's first component and the sub bytes in the AES algorithm of encryption, including two mathematical transformations: the Finite Galois Field GF (2^8) with a multiplicative inverse g and an invertible affine transformation f. The transformation possesses byte non-linearity.

In S(p_old)= f (g(p_old)): p_old -> p_ new, new substitution byte p_new replaces the every old byte p_old. The g and f functions are applied to every number in [0×00; 0×ff] range for obtaining Substitution Box mapping (S) among every value present in [0×00; 0×ff]. Through the mapping of s-box, p_new output is obtained instantly and simply from every input of s-box's p_old in the block cipher. The position of the input is permuted by the process of modified Bernoulli, which is motivated toward the diffusion effect reinforcement. At last, from the modified Bernoulli process, output blocks are obtained, the pixel between them is applied with the operator of XOR, and Kc, the dynamical key, fulfills the task of masking. PCNG feeds Kp and Kc and continuously changes their values for every block and round r. A round of every block uses 8 bits in Kp, implementing 1/4 of the PCNG samples (XðnÞ) and since Kc requires 4 PCNG samples with 16 bytes, $1/4^{th}$ of PCNG samples is required to encrypt a block. PCNG samples of $[(1/4 + 4) \times r \times N_b]$ are required for a block cipher, where the processed text with several blocks is N_b.

ALGORITHM

Step 1: Establish A as integer parameter thus $A > 0$ and $A \neq k \times 257, k = 1, 2, 3, ...$

Step 2: For a T array, consider $T \leftarrow [0, 1, 2, ..., 255]$, $[0, 255]$ is the range with 256 distinct integers of array T.

Step 3: R is the new array obtained from A and T and the below linear mapping:

$$R(i) = \mod((A \times (T(i) + 1)), 257), i = 1, 2, ..., 256$$

The output obtained is which is not exactly divided by 257. Like, . where positive integer is represented as A satisfies, and positive integer is given by k. is not considered an integer, and is also not considered an integer. Therefore, $T(i) \in \{0, 1, 2, ..., 255\} \rightarrow R(i) \in \{1, 2, ..., 256\}$.

Step 4: consider $R(i) \leftarrow R(i) - 1$ so $R(i) \in \{0, 1, ..., 255\}, i = 1, 2, ..., 256$. The array $\mathbf{R} = \{R(i)\}$ in 1D is obtained.

Step 5: The 1D R array is transformed into an Rb 2D matrix and considers the initial s-box Rb.

Step 6: Establish the μ parameter, x_0 as the tent–logistic map's initial state value, and L as an integer which is highly greater than 256. After that tent, the logistic map is iterated by L times to generate a Chaos sequence with L length. The output chaotic sequence sensitivity is enhanced to its initial state value. (L-256) is the original elements present in the sequence of Chaos, with the sequence length of 256 and denoted by X.

Step 7: X chaotic sequence is sorted, obtaining $J = \{J(1), J(2), ..., J(256)\}, J(i) \in \{1, 2, ..., 256\}$ an array of position index. Because of the chaotic sequence's ergodicity and non-periodicity, inevitable to obtain $J(i) \neq J(j)$ till $i \neq j$.

Step 8: Compute the S1 1D array as:

$$S1(i) = T(J(i)), i = 1, 2, ..., 256.$$

Step 9: Convert the S1 into Sb.

6.5 Performance Analysis

The chaos-based advanced encryption algorithm using the S-box Bernoulli process (CA-Sbox-BP) designed here has been captured with VHDL as well as designs instantiated with Spartan-3 XC3S400, Spartan-3 XC3S50 (iterative), Spartan-3 XC3S50 (serial), Virtual-5 XC5VLX50(iterative), and Virtex-5 XC5VLX50(serial). This design is synthesized with FPGA devices by comparing existing methods such as the chaos-based affine

TABLE 6.1

Simulation Settings of Various FPGAs

Platform	Flipflops	LUT	Slices
Spartan-3 XC3S400	300	478	203
Spartan-3 XC3S50(iterative)	178	325	142
Spartan-3 XC3S50 (serial)	-	-	156
Virtex-5 XC5VLX50(iterative)	564	156	226
Virtex-5 XC5VLX50(serial)	271	321	54

transformation generation (CATG) method and linear congruence generator s-box (LCGS-box). The parameters, namely throughput, latency, efficiency, non-linear criterion, and avalanche effect, were employed to estimate the proposed CA-Sbox-BP implementation. To implement the proposed approach, simulation parameter settings are presented in Table 6.1.

- **Efficiency**
  The efficiency of S-boxes is stated to reduce overhead time and maintain good accuracy for different slides. It can be defined as follows:

$$\text{Efficiency}\,(\%) = \frac{\text{throughput}}{\text{number of utilized slices}} \tag{6.5}$$

Table 6.2 shows the comparison of efficiency between existing chaos-based affine transformation generation (CATG) method and linear congruence generator s-box (LCGS-box) with proposed chaos-based advanced encryption algorithm using the S-box Bernoulli process (CA-Sbox-BP)

Figure 6.3 shows the efficiency comparison between various FPGA platforms by analyzing existing CATG, LCGS-box, and the proposed CA-Sbox-BP. It is found that the existing methods such as CATG and LCGS-box achieve 92% and 94%, and hence, the proposed method CA-Sbox-BP achieves 96%, which is better than CATG and LCGS-box as improved by 4% and 2%, respectively. In contrast, the X-axis shows various s-box, and the Y-axis shows the efficiency values obtained in percentage.

- **Throughput**
  With several stages and an additional register count, the number of blocks per second increases linearly through this throughput.

TABLE 6.2

Analysis of Efficiency

FPGA Platforms	CATG	LCGS-box	CA-Sbox-BP
Spartan-3 XC3S400	87	89	90
Spartan-3 XC3S50(iterative)	89	90	92
Spartan-3 XC3S50 (serial)	91	93	95
Virtex-5 XC5VLX50(iterative)	93	95	96
Virtex-5 XC5VLX50(serial)	95	96	97

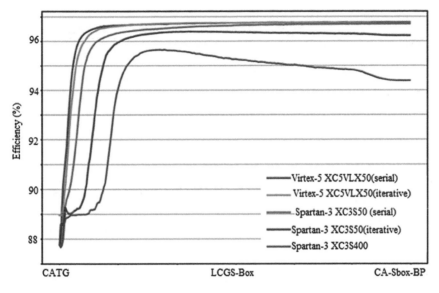

FIGURE 6.3
Comparison of efficiency.

TABLE 6.3

Analysis of Throughput

FPGA Platforms	CATG	LCGS-box	CA-Sbox-BP
Spartan-3 XC3S400	88	90	92
Spartan-3 XC3S50(iterative)	90	91	94
Spartan-3 XC3S50 (serial)	92	94	95
Virtex-5 XC5VLX50(iterative)	95	96	96
Virtex-5 XC5VLX50(serial)	96	97	98

Table 6.3 gives a comparison between throughput with existing chaos-based affine transformation generation (CATG) method and linear congruence generator s-box (LCGS-box) with proposed chaos-based advanced encryption algorithm using the S-box Bernoulli process (CA-Sbox-BP)

Figure 6.4 shows the throughput between various FPGA platforms by analyzing existing CATG, LCGS-box, and the proposed CA-Sbox-BP. In contrast, the X-axis shows various s-box, and the Y-axis gives throughput values in kbps. These results indicate that methods such as CATG and LCGS-box achieve 93 kbps and 95 kbps, respectively, and the proposed method CA-Sbox-BP achieves 98 kbps, which is better than CATG and LCGS-box improved by 5 kbps and 3 kbps, respectively.

- **Latency**
  It is about the minimum processing time for one block independently of other blocks. Table 6.4 gives a comparison between latency and existing chaos-based affine transformation generation (CATG) method and linear congruence generator s-box (LCGS-box) with proposed chaos-based advanced encryption algorithm using the S-box Bernoulli process (CA-Sbox-BP).

  Figure 6.5 shows the latency comparison between various FPGA platforms by analyzing existing CATG, LCGS-box, and the proposed CA-Sbox-BP. It is found that the

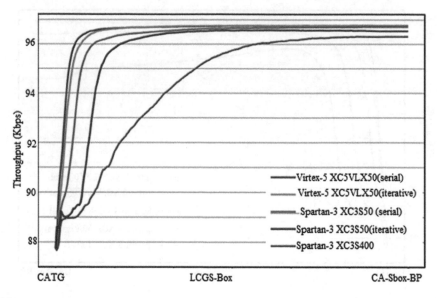

FIGURE 6.4
Comparison of throughput.

TABLE 6.4

Analysis of Latency

FPGA Platforms	CATG	LCGS-box	CA-Sbox-BP
Spartan-3 XC3S400	80	78	75
Spartan-3 XC3S50(iterative)	82	80	78
Spartan-3 XC3S50 (serial)	85	82	80
Virtex-5 XC5VLX50(iterative)	86	85	81
Virtex-5 XC5VLX50(serial)	87	86	82

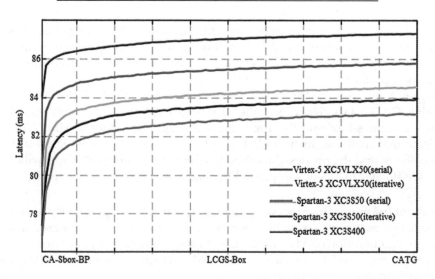

FIGURE 6.5
Comparison of latency.

existing methods such as CATG and LCGS-box achieve 81 ms and 79 ms, respectively, and hence the proposed method CA-Sbox-BP achieves 75 ms, which is better than CATG and LCGS-box as improved by 6 ms and 4 ms, respectively. In contrast, the X-axis gives s-box, and Y-axis gives latency obtained in ms.

- **Non-linear criterion**
  Non-linear criterion is a predominant characteristic in S-boxes performance evaluation system. The higher non-linearity gives a stronger ability in S-box for preventing non-linear attacks. The non-linear criterion is given as follows:

$$Nf = \min[dH(f, l)]$$

Table 6.5 gives a comparison between non-linear criterion with existing chaos-based affine transformation generation (CATG) method and linear congruence generator s-box (LCGS-box) with proposed chaos-based advanced encryption algorithm using the S-box Bernoulli process (CA-Sbox-BP).

Figure 6.6 shows the comparison of non-linear criteria between various FPGA platforms by analyzing existing CATG, LCGS-box, and the proposed CA-Sbox-BP. It

TABLE 6.5

Analysis of Nonlinear Criterion (%)

FPGA Platforms	CATG	LCGS-box	CA-Sbox-BP
Spartan-3 XC3S400	70	68	65
Spartan-3 XC3S50(iterative)	72	70	68
Spartan-3 XC3S50 (serial)	75	72	70
Virtex-5 XC5VLX50(iterative)	76	75	71
Virtex-5 XC5VLX50(serial)	77	76	72

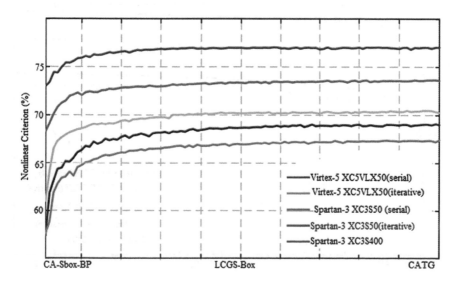

FIGURE 6.6
Comparison of non-linear criterion.

TABLE 6.6

Analysis of Strict Avalanche Effect

FPGA Platforms	CATG	LCGS-box	CA-Sbox-BP
Spartan-3 XC3S400	50	48	45
Spartan-3 XC3S50 (iterative)	52	50	48
Spartan-3 XC3S50 (serial)	55	52	50
Virtex-5 XC5VLX50 (iterative)	56	55	51
Virtex-5 XC5VLX50 (serial)	57	56	52

is found that the existing methods such as CATG and LCGS-box achieve 65% and 67%, and hence, the proposed method CA-Sbox-BP achieves 64%, which is better than CATG and LCGS-box as improved by 1% and 3%, respectively. In contrast, the X-axis gives s-boxes, and the Y-axis gives non-linear criteria obtained in percentage.

- **The strict avalanche effect**
  The SAC correlation matrix can measure the effect of S-boxes. The S-box satisfies the strict avalanche effect if each sac correlation matrix is close to 0.5.
  Table 6.6 shows the comparison of strict avalanche effect between existing chaos-based affine transformation generation (CATG) method and linear congruence generator s-box (LCGS-box) with proposed chaos-based advanced encryption algorithm using the S-box Bernoulli process (CA-Sbox-BP).

  Figure 6.7 compares the strict avalanche effect between various FPGA platforms by analyzing existing CATG, LCGS-box, and the proposed CA-Sbox-BP. In contrast, the X-axis shows various s-box, and the Y-axis gives a strict avalanche effect in percentage. Thus, existing methods such as CATG and LCGS-box achieve 45% and 43%, and hence, the proposed method CA-Sbox-BP achieves 41%, which is better than CATG and LCGS-box as improved by 4% and 2%, respectively. Table 6.7 shows the

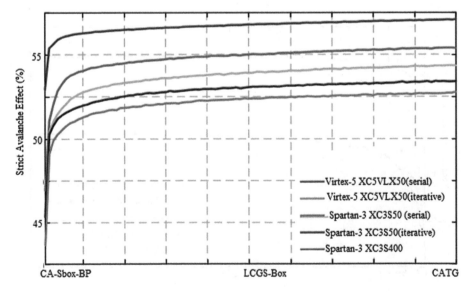

FIGURE 6.7
Comparison of the strict avalanche effect.

TABLE 6.7

Overall Comparison Between Existing and Proposed Methods

Parameters	CATG	LCGS-box	CA-Sbox-BP
Efficiency (%)	92	94	96
Throughput (kbps)	93	95	98
Latency (ms)	81	79	75
Non Linear Criterion (%)	65	67	64
Strict Avalanche Effect (%)	45	43	41

overall comparison between the existing chaos-based affine transformation generation (CATG) method and linear congruence generator s-box (LCGS-box) with proposed chaos-based advanced encryption algorithm using the S-box Bernoulli process (CA-Sbox-BP).

6.6 Summary

A secure, robust cryptosystem depends on AES substitution box, and chaotic components are proposed in this chapter. An effective PCNG, global diffusion, and block cipher are comprised. The finite field is used to define the PCNG, which eliminates the danger of depreciated security as a subsequent dynamic degradation. At the same time, numerical implementation of chaotic maps is always stated as real numbers. In the modified Bernoulli process, diffusion properties are increased efficiently by HAD and VAD. The strength of cryptography is tested in detail in S-Box proposed through the various benchmark standards. The results of experiments and examination of performance showed that this chaos-based advanced encryption algorithm proposed using the S-box Bernoulli process (CA-Sbox-BP) achieves 96% efficiency, 64% non-linear criterion, and 41% strict avalanche effect with 98 kbps in 75 ms. The future work concentrates on including chaotic logistic and tent maps for the effective area and power reduction in the s-box.

References

H. A. Ahmed, M. F. Zolkipli and M. Ahmad, "A novel efficient substitution-box design based on firefly algorithm and discrete chaotic map", *Neural Computing and Applications*, 2019. DOI: 10.1007/s00521-018-3557-3

A. Belazi, M. Khan, A. A. A. El-Latif and S. Belghith, "Efficient cryptosystem approaches S-boxes and permutation–substitution-based encryption", *Nonlinear Dynamics*, vol. 87, pp. 337–361, 2018.

A. K. Farhan, R. S. Ali, H. Natiq and N. M. G. Al-Saidi, "A new S-box generation algorithm based on multistability behaviour of a plasma perturbation model", *IEEE Access*, vol. 7, pp. 124914–124924, 2019.

J. M. Guo, D. Riyono and H. Prasetyo, "Improved beta chaotic image encryption for multiple secret sharing", *IEEE Access*, vol. 6, pp. 46297–46321, 2018.

U. Hayat, N. A. Azam and M. Asif, "A method of generating \$8times8\$ substitution boxes based on elliptic curves", *Wireless Pers. Commun.*, vol. 101, no. 1, pp. 439–451, Jul. 2018.

R. Hosseinkhani et al., "Using cipher key to generate dynamic S-Box in AES cipher system", *International Journal of Computer Science and Security (IJCSS)*, vol. 6, pp. 19–28, 2012.

I. Hussain, T. Shah, M. A. Gondal and H. Mahmood, "Efficient method for designing chaotic S-boxes based on generalized Baker's map and TDERC chaotic sequence", *Nonlinear Dynamics*, vol. 74, no. 1, pp. 271–275, Oct. 2013.

L. Jingmei, et al., "One AES S-box to increase complexity and its cryptanalysis", *Journal of Systems Engineering and Electronics*, vol. 18, no. 2, pp. 427–433, 2007.

M. F. Khan, A. Ahmed, K. Saleem and T. Shah, "A novel design of cryptographic SP-network based on gold sequences and chaotic logistic tent system", *IEEE Access*, vol. 7, pp. 84980–84991, 2019.

L. Kocarev, "Chaos-based cryptography: A brief overview", *IEEE Circuits Syst. Mag.*, vol. 1, no. 3, pp. 6–21, Mar. 2001.

Dragan Lambić, "A new discrete-space chaotic map based on the multiplication of integer numbers and its application in S-box design", *Nonlinear Dynamics*, vol. 100, no. 1, pp. 699–711, 2020.

Q. Lu, C. Zhu and X. Deng, "An efficient image encryption scheme based on the LSS chaotic map and single S-box", *IEEE Access*, vol. 31, no. 8, pp. 25664–25678, Jan. 2020.

Q. Lu, C. Zhu and G. Wang, "A novel S-box design algorithm based on a new compound chaotic system", *Entropy*, vol. 21, no. 10, p. 1004, Oct. 2019.

S. Murugan, S. Jeyalaksshmi, B. Mahalakshmi, G. Suseendran, T. N. Jabeen and R. Manikandan, Comparison of ACO and PSO algorithm using energy consumption and load balancing in emerging MANET and VANET infrastructure. *Journal of Critical Reviews*, vol. 7, no. 9), 2020. DOI: 10.31838/jcr.07.09.219

F. Özkaynak, "Brief review on application of non-linear dynamics in image encryption", *Nonlinear Dynamics*, vol. 92, pp. 305–313, 2018.

R. Paul, S. Saha, J. K. M. S. U. Zaman, S. Das, A. Chakrabarti and R. Ghosh, A simple 1-byte 1-clock RC4 hardware design and its implementation in FPGA coprocessor for secured ethernet communication, *Proc. National Workshop on Cryptology*, VIT University & CRSI, Vellore, India, Aug. 6–8, 2012.

P. Ping, J. Fan, Y. Mao, F. Xu and J. Gao, "A chaos-based image encryption scheme using digit-level permutation and block diffusion", *IEEE Access*, vol. 6, pp. 67581–67593, 2018.

R. Rahim, S. Murugan, S. Priya, S. Magesh, and R. Manikandan, "Taylor based Grey Wolf Optimization Algorithm (TGWOA) for energy-aware secure routing protocol", *International Journal of Computer Networks And Applications*, vol. 7, no. 4, p. 93, 2019.

G. Tang, X. Liao and Y. Chen, "A novel method for designing S-boxes based on chaotic maps", *Chaos Solitons Fractals*, vol. 23, no. 2, pp. 413–419, Jan. 2005.

E. Tanyildizi and F. Ozkaynak, "A new chaotic S-box generation method using parameter optimization of one-dimensional chaotic maps", *IEEE Access*, vol. 7, pp. 117829–117838, 2019.

X. Wang, X. Zhu and Y. Zhang, "An image encryption algorithm based on Josephus traversing and mixed chaotic map", *IEEE Access*, vol. 6, pp. 23733–23746, 2018.

L. Yi, X. Tong, Z. Wang, M. Zhang, H. Zhu and J. Liu, "A novel block encryption algorithm based on chaotic S-box for wireless sensor network", *IEEE Access*, vol. 7, pp. 53079–53090, 2019.

T. Zhang, P. Chen, L. Chen, X. Xu and B. Hu, "Design of highly non-linear substitution boxes based on I-Ching operators", *IEEE Transactions on Cybernetics*, vol. 48, no. 12, pp. 3349–3358, Dec. 2018.

S. Zhu, C. Zhu and W. Wang, "A novel image compression-encryption scheme based on chaos and compression sensing", *IEEE Access*, vol. 6, pp. 67095–67107, 2018.

C. Zhun and K. Sun, "Cryptanalyzing and improving a novel colour image encryption algorithm using RT-enhanced chaotic tent maps", *IEEE Access*, vol. 6, pp. 18759–18770, 2018.

7

Cyber-Physical Systems Attacks and Countermeasures

Philip Asuquo, Midighe Usoh, Bliss Stephen, Aneke chikezie Samuel, and Afolabi Awodeyi

University of Uyo, Akwa Ibom State, Nigeria

CONTENTS

7.1 Introduction

Cyber-physical systems (CPS) have been identified as critical components of the industrial internet of things (IIoT) and are expected to play a significant part in Industry 4.0. CPS provides accurate and real-time operation of intelligent applications and services (Gupta et al., 2020). CPS has grown in popularity in many fields of human endeavor, especially where physical processes and equipment must be coordinated with persons, systems, or subsystems. New concepts like Industry 4.0 (Wang et al., 2011) and Industrial Internet (Amin et al., 2013) will require increased automation, autonomy, and a new understanding of industrial processes. The requirement for high degree of communication between components and abstraction of physical processes is driving the development of CPS (Hamdan et al., 2021). CPS have a variety of characteristics, including the ability for discrete components to communicate with one another, resulting in complex systems (Aazam et al., 2018). Data acquisition in CPS is achieved using sensor devices which transmit data using networks to the control systems when there are no interactions between human and machines in some instances (Ashibani and Mahmoud, 2017). Sensors, barcodes, and radio-frequency identification (RFID) tags are rapidly being embedded in physical objects so that they can be scanned by smart devices. These devices can be connected to the Internet in order to transmit identifiable data and location information for the purpose of monitoring

DOI: 10.1201/9781003241348-7

and managing the physical environment (Juma and Shaalan, 2020). In addition, the computing and processing units can be located on the cloud, with the resulting decisions being communicated to physical objects via actions (Yemini et al, 2020). Early in the design process, security considerations relating to the type of technology used and the development framework should be considered (Howladar et al., 2021). In addition, as a result of the inherent characteristics and benefits of available networks such as wireless sensor networks (WSN), next-generation networks (NGN), and the Internet, CPS are increasingly exposed to new security challenges, such as securing protocols and trusting relationships between CPS components (Chawla et al., 2021).

Information security, as defined by ISO/IEC 27001 (Wang et al., 2011), is the process of ensuring the confidentiality, integrity, and availability of information. Threats and vulnerabilities become more difficult to assess, and new security challenges occur as a result of complex cyber-physical interactions. In addition, identifying, tracing, and examining threats that originate from, move between, and target numerous CPS components is challenging (Islam et al., 2015). It is critical to have a thorough awareness of vulnerabilities, threats, and assaults before developing security systems (Khaitan and McCalley, 2015). This chapter presents the current state of the art in CPS security, vulnerability assessment, threats, and mitigation schemes. An examination of the security requirements and issues associated with the CPS as well as a discussion of potential solutions and opportunities for future research is presented. The rest of the proposed chapter will be organized as follows. In Section 7.2, the current state of the art and security requirement for CPS will be presented. In Section 7.3, we will discuss the different types of threats and attacks in CPS. Different control mechanisms for mitigation of threats will be discussed in Section 7.4. We outline the major threats, vulnerabilities, attacks, and control methods in each CPS application domain in Section 7.5. Open issues and security challenges in CPS will be discussed in Section 7.6. The proposed book chapter will be completed in Section 7.7 with summary of findings on CPS security.

7.2 CPS Requirements

The physical world is monitored and controlled by CPS. It should be built to meet the following requirements in order to meet real-world constraints.

- **Response in real time**: Cyber-physical systems are expected to meet real-time constraints depending on the application. When it comes to measurement data, for instance, if the process is electrical, the system response time ought to be better than water systems. Each process, however, has its own set of real-time performance requirements. Any delay in signal propagation caused by a defect or an exploit (e.g., a denial-of-service attack) can be catastrophic.

- **Limitation of resources**: It is well known that CPS has a lot of resource-constrained devices. From analog to digital converters, sensing devices, input and output devices that operate remotely, and controllers for example are often developed to carry out specific functions with the limitations they have in terms of storage and computational capabilities. The main objective is to have devices that are very robust, to operate over an extended period of time and to meet real-time performance constraints.

- **Availability**: It is a lot more difficult to shut down a plant than it is to restart a server. The availability of a CPS is a crucial prerequisite. The sensitive nature of these systems necessitates extremely high availability, like in the instance of a temperature regulator in a critical biological process. As a result of the high uptime, changing hardware and software is especially difficult for CPSs. The main goal is to avoid interfering with the CPS's functionality.

7.3 CPS as a Critical Infrastructure

In recent times, CPS have replaced traditionally engineered systems in the electricity, mobility, healthcare, and manufacturing sectors. These CPS are mission critical in general: their reliability and efficient operation are vital. This section explores categories of CPS as a critical infrastructure:

1. **Security-aware applications**: A cybersecurity threat on a security-aware application often leads to CPS security breach from security flaws such as rootkits, poorly implemented codes, and backdoors.

2. **Mission-critical applications**: Mission-critical systems are those that are required for the operation and survival of a business or organization to function. A mission critical system failure or interruption has a major impact on the ability of a company to conduct its business.

3. **Business-critical applications**: Any system on which an organization relies in order to carry out the routine business operations that are necessary in order for the organization to function successfully is referred to as a business-critical system. To put it another way, when mission-critical systems are required for the success of a business, they are referred to as business-critical systems. In the event that a mission-critical system is under attack, businesses may suffer financial losses, customer discontent, and productivity losses.

4. **Safety-critical applications**: It is necessary for a safety essential system to operate properly in order to prevent human injury and death, property destruction and loss of revenue, as well as environmental degradation and severe systemic implications. When the use of a system is associated with a danger, it is said to be safety critical (a serious mistake which could have devastating effects). Safety-critical systems are designed to ensure the safe operation of systems that are subject to hazard, which is defined as any state or condition under which the system's improper operation will inevitably result in a mishap; for example, a train traveling at a high rate of speed is considered hazardous.

7.4 CPS Architecture

A CPS can be made up of numerous sensors and actuators that are linked together by an intelligent decision system (Kumar and Patel, 2014) as shown in Figure 7.1. CPSs exhibit characteristics such as cross-domain sensor collaboration, heterogeneous flow of

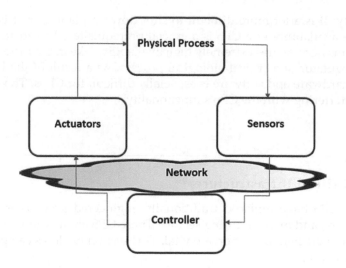

FIGURE 7.1
Architecture of a cyber-physical system.

information, and intelligent decision-making. At different levels, the CPS architecture can be considered. The most frequently used architecture for CPS is divided into seven main layers of the ISO/OSI framework (Johnson, 2020; Kocabas et al., 2016).

1. **Physical layer**: Embedded sensors, actuators, terminals of mobile devices, and RFID-based tag readers comprise the physical layer (Ferdinando et al, 2006). These elements link the physical world to the physical layer. These components' primary job is to monitor or track by collecting data, sound, heat, and executing pre-programmed orders. Once data have been collected and sent to an application, they are examined and processed to allow for decision-making. Modifying or tampering with these data at this level could have system-wide disastrous repercussions. As a result, a communication code that is secure or an encryption mechanism is required to ensure adequate privacy protection. One of the problems at this layer is the retention of the secret key by the controller after decoding (Kumar and Patel, 2014). What is required is a type of encryption known as homomorphic encryption, which allows calculations to be performed on data that are encrypted without first decrypting the data. Data must be accurate and secured in this manner within the application (Johnson, 2010).

2. **Network layer**: The network layer's principal duty is to convey and exchange massive volumes of data as well as signal command control between the application and the physical layer. The next generation of networks, such as the internet, local area network, Bluetooth, and Wi-Fi, manage the network layer. When a huge amount of data and transformation surpasses the layer's capabilities, there may be an issue. This can cause network congestion, making it vulnerable to denial-of-service (DOS) assaults, distributed denial-of-service (DDOS) attacks, and black-hole attacks. Data loss integrity attacks can be caused through deception attacks (Ye et al., 2016).

3. **Application layer**: The application layer is the primary component of the CPS where intelligent environments are developed. Smart grid functions are placed

with other functions to be completed within intelligent settings. This layer's first task is to receive and analyze execution control commands. From here, all commands are transferred, and complicated algorithms are used to do their procedures on the collected data (Ferdinando et al, 2006). Another task on this layer is to monitor the overall cyber-physical system. The application layer monitors system behavior and alters physical equipment behavior to govern what is preprogrammed for (Ferdinando et al, 2006). The storage system can also be placed under this layer to control any event that modifies the CPS's behavior. The most crucial role of this layer is to protect data privacy (Chun et al., 2010). Any infringement of privacy here, such as private data leakage, malicious code, and structured query language (SQL) injection assaults, would be identified as a threat to the entire system.

7.5 An Overview of Security Challenges in Cyber-Physical Systems

In today's digital era, cybersecurity concerns are common, and new cyber incidents are reported on a regular basis. In reality, cyberattacks may have had a direct impact on a large number of people (Heng, 2014). Most significantly, a cybersecurity attack that was targeted at a retail store (Gries et al., 2017) harmed up to one-third of the population of the United States. Credentials that were stolen from a targeted vendor were used to exploit the system in this case (Ferdinando, 2006). The type of cyberattack that harmed target and its customers is only one illustration of the many ways cyberattacks may be carried out. Smaller breaches are nonetheless costly, costing $5.4 million in 2012, whereas the average cost of data theft in the United States in 2012 was $188 per user account (Chun et al., 2010). As indicated by public data, there has been a noticeable surge in CPS breaches. In 2013, the typical American corporation was subjected to 16,856 attacks (Rad et al., 2015). In 2014, the number of reported data breaches and cyberattacks in the industry increased by 23.9 percent over 2013, with 761 breaches exposing 83,176,279 records (Haidegger et al., 2020). These cyberattacks are damaging for individuals and the government as a whole. More importantly, data network and computer systems, as well as their associated distribution channels, are key components of our nation's vital infrastructure.

Securing CPSs involves the implementation of security and resilience policies, as well as their application in system design and implementation. To effectively implement services that enhance our knowledge of various interdependencies, forecast future conditions, and aid in decision- and policy-making, methods and tools from system and game theory, appropriate techniques, predictive analysis, incentive technology, sociology, and cognitive science must be combined (Siddappaji and Akhilesh, 2020). In many domains where CPS have been effectively deployed (Lee et al., 2015), such as controlling chemical processes and automating buildings, the value of CPS has grown markedly. As a result of these achievements, the CPS control system and network are subject to real-time attacks (Chun et al., 2010). Hackers could, for example, deny service by utilizing connections from control systems, which will result in delays in responses and, as a result, serious system-wide problems with power grid infrastructure (Lee et al., 2015). Signal encryption could be an efficient way to improve security in situations where continuous, dependable transformation functionality via the network is required. However, using encryption alone might not be efficient in the mitigation risks of disclosure by insiders or delays caused by inefficient algorithms (Lee et al., 2015), widespread system overload with a growing number of devices that interact (Rad et al., 2015).

7.6 Security Challenges in CPS

Adopting security measures provides numerous advantages such as securing the components of CPS, levels, and application domains. Regardless of these advantages, there is an influence on the CPS system whenever these cryptographic mechanisms are implemented, which can be stated as follows:

1. **Delays due to operational procedures**: When any type of security solution is implemented, a training occurs prior to when security mechanisms are enabled, during which the service is unavailable or rudimentary and hence vulnerable to cybersecurity threats.

2. **High computational power**: CPS end-devices with little computational capabilities and energy-constrained face a major problem, particularly in this case. The lifespan of the device is shortened when the power consumption is high and a higher cost of operation is required to keep them up and running.

3. **Transmission latency**: In real-time and critical systems, cryptographic protocols such as encryption mechanisms incur additional transmission overhead which result in very high computational delays and packet losses. Although these security mechanisms provide solutions to mitigate CPS attacks such as active and passive eavesdropping, man-in-the middle attacks and several other attacks can distort transmission of data from source to destination CPS devices.

   Despite the fact that it provides a defensive advantage, it is very unsuitable for a real-time critical protection system.

4. **Computational overhead**: Standard cryptographic protocols incur additional overhead cost to data transmission. Traditional security mechanisms such as public key infrastructure (PKI), which is asymmetric, add more computational cost in CPS networks. These cryptographic mechanisms incur additional overhead from the key initialization phase, training, update, and operating stages, and when there is a need for certificate revocation due to a malicious node in the CPS network.

5. **Performance reduction**: The performance of CPS systems can be impaired when ubiquitous and traditional cryptographic security mechanism are implemented. For real-time and time and event applications, this degradation can disrupt routine or event-driven operations which may require manual interventions to keep services up-to-date.

6. **Interoperability issues**: One of the main issues in CPS system is interoperability. Technical, syntactic, semantic, and cross-domain interoperability are the four layers of interoperability established by the European Telecommunication Standards Institute (ETSI) and European Interoperability Framework (EIF) for complex systems. Because of the complex nature of CPS, many security mechanisms are not compatible and cannot be implemented on these systems. This could be due to firmware or the operating system in use. The security challenges in CPS are depicted in Figure 7.2.

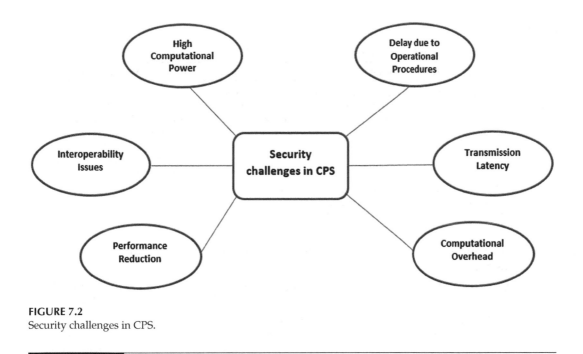

FIGURE 7.2
Security challenges in CPS.

7.7 Security Threats and Vulnerabilities in Cyber-Physical System

The threats in CPS are classified into different layers and they are:

1. **Physical Layer Attacks**

   The physical layer attacks are jamming DoS and node capture.

   a. **Jamming DoS**: This is a type of DoS is implemented at the physical layer (Hu and Sharma 2005). Signals can be jammed by a malicious device through transmitting at similar frequency. This promotes noise in the carrier and can decrease the signal-to-noise ratio to the level whereby the nodes cannot receive correct data. Jamming can prevent the nodes from communicating. Another way of preventing transmissions is by implementing jamming temporarily at random intervals.

   b. **Node capture (tampering)**: This is when an attacker physically gains control of a sensor node such as connecting a cable to a network and having access to an ongoing transmission in a WSN (Butun et al., 2013). This can enable the attacker to tamper data stored in a memory of a node and control the node. Node capturing is very critical as cryptography-related keys can be exposed, leading to the adversary having control over the WSN. The problem here is that these compromised nodes can either transmit frivolous queries (availability attack) or supply incorrect data to the authorized users (integrity attack). Table 7.1 gives a summary of security threats, mode of attack, result of attack, and the countermeasure of attacks in physical layer.

2. **Data Link Layer Attacks**

   The data link is susceptible to DoS attack because of the data link algorithms, specifically medium access (MAC) schemes. A channel may be consistently jammed by a MAC DoS attack which is a typical example of presenting a window for DoS

TABLE 7.1

Summary of Physical Layer Threats in Cyber-Physical System

OSI Layer	Security Threat	Mode of Attack	Result of Attack	Countermeasure
Physical layer	Jamming DoS	Using malicious node through similar frequency	Prevents nodes from receiving legitimate attack	Detect and sleep, routing around jammed areas, simply hiding the nodes by camouflaging, spread-spectrum, message priority, reduced duty cycle, area mapping, mode change Tamper-proofing
	Node capture (Tampering)	Physically connecting cable to the node	Tamper stored data in memory	Tamper-proof boxing

attack (Butun, Österberg, and Song 2020). The following attacks can be executed in the data link layer:

a. **Collision**: An attacker monitors when the authorized node in the network is sending packets and transmits its own packet simultaneously from the same authorized node channel, thus leading to collision of packets. This prevents the receiver from obtaining a complete meaningful packet from the transmitter owing to the fact that packet collision leads to collision loss during transmission (Cyclic Redundancy Check) (Borgohain, Kumar and Sanyal, 2015).The receiver disposes the packets because it is useless and sends a request to the transmitter to re-send the packets again. Attackers prefer collision attack more than jamming attack because the energy transmitted and the possibility of detecting it is lower (Znaidi, Minier and Babau, 2008).

b. **Sleep deprivation**: Collision attacks or repeated handshaking (request to send and clear to send) can prevent the node from sleeping. These cause the battery to drain its stored energy. The aim of this attack is to compel the node to exhaust all the energy in its batteries (Butun et al., 2013).

c. **Desynchronization**: Time synchronized channel hopping (TSCH), which is a MAC layer protocol, can be attacked by transmitting data in the scheduled time allocated to other users (Sajjad and Yousaf, 2014). This leads to collision and loss of packet. When the attacker generates a series of these events, the neighboring motes becomes de-synchronized.

d. **Exhaustion**: When the chosen node of attack drains its battery energy due to desynchronization, this attack is referred to as exhaustion attack (Znaidi, Minier and Babau, 2008). Laptops or a simple node with the capacity to send radio signals using similar sensors' band under attack can be used to implement this kind of attack (Butun, Österberg and Song, 2019).

e. **Link layer flooding**: In link layer flooding, the neighboring nodes of the compromised nodes receive unnecessary data packets or controlled packets from MAC. The aim of this attack is to drain battery power or execute DoS. Furthermore, link layer flooding can also lead to the depletion of channel bandwidth resources (Liu, Li, and Man, 2005).

f. **Link layer jamming**: This attack focuses on jamming important packets such as data packets. The arrival time of the data packet is retrieved and is used against the transmission of packets. MAC protocols such as B-MAC, L-MAC, and S-MAC are susceptible to link layer jamming (Znaidi, Minier and Babau, 2008).

g. **Spoofing/ARP-spoofing**: The MAC address of a victim node is imitated by the malicious node. The malicious nodes then use the MAC address to generate different legitimate identities and uses these identities to execute different tasks in the network (Shabana, Fida, Khan, Jan, and Rehman, 2016). In ARP-spoofing, the spoofed address resolution protocol (ARP) is transmitted into the network. The goal of the attacker is to divert the traffic meant for legitimate node to the attacker.

h. **Unfairness**: Unfairness to the network can be generated by an attacker implementing exhaustion attack or exploiting cooperative MAC protocols periodically (Wood, and Stankovic, 2002). The node in the network experiences occasional blackouts, thereby causing delay in the sending/receiving of messages. The quality of service in this network is degenerated causing the network to lose out on the allocated real-time MAC protocol configuration transmission deadlines. Table 7.2 summarizes the security threats, mode of attack, result of attack, and the countermeasure of attacks in datalink layer.

3. **Network Layer Attacks**
The network layer is attacked by sending a huge number of packets into the network. This will cause the network traffic to be congested and also causes power resources in the network to be denied (Yang, Luo, Ye, Lu, and Zhang, 2004).The following are the network layer attacks:

a. **HELLO flooding**: The attacker first of all gains a longer range in transmission and then transmits a broadcast advertisement message to the entire network. The nodes in the network are convinced that the broadcast messages are sent from a neighboring node ((Yang, Luo, Ye, Lu, and Zhang, 2004). Once these legitimate nodes transmit packets to the malicious node, other legitimate nodes cannot receive any packet (Karlof, and Wagner, 2003).

b. **Hole attacks**

- **Blackhole**: In blackhole, all the packets intended for forwarding is discarded by a malicious node. Blackhole attack, which can also be known as "selfishness", is potent when the blackhole is also a sinkhole. The traffic in the blackhole may be terminated by the combination of these attacks (Kaur and Singh, 2014).

- **Sinkhole**: When a malicious node pretends to be the most appropriate node for transmitting packets to its destination, this is referred to as a sinkhole. The malicious node turns into a hub for receiving packets. Unlike blackhole, sinkhole does not discard the packets, thus becoming undetected (Kibirige and Sanga, 2015).

- **Selective forwarding (grayhole)**: This is another exceptional modification of blackhole attack whereby the malicious node selects the received packets to be discarded. Just like sinkhole attacks, the malicious nodes corrupt the routing protocol and allocates many routes to itself only to drop selected packets that it receives (Brar and Angurala, 2016).

- **Wormhole**: Two nodes located in remote parts of the network can create a tunnel to enable them transmit packets faster (Hu et al., 2003). The malicious node can either snoop around or obtain the data packets and transmit

TABLE 7.2

Summary of Datalink Layer Threats in Cyber-Physical System

OSI Layer	Security Threat	Mode of Attack	Result of Attack	Countermeasure
Datalink layer	Collision	Forged ARP packets are used to flood the switch with MAC addresses until the content addressable memory (CAM) table is full	Prevents receiver from obtaining complete meaningful packets	Authentication and anti-replay protection, decreasing the MAC admission control rate, employing time division multiplexing (TDM) technique, error-correcting code
	Denial of Sleep	Collision attacks or repeated handshaking	Exhaust node energy	Authentication mechanisms and anti-replay strategies are used for mitigation, detect and sleep, broadcast attack protection
	Desynchronization	Attacking time synchronized channel hopping (TSCH)	Collision and loss of packets	Configure switches to limit ports with DHCP requests capacity, implement static ARP, secure communication of MAC layer frames
	Exhaustion	Forged ARP packets are used to flood the switch with MAC addresses	Targeted receiver does not receive the whole meaningful packet from the transmitter	Limit the MAC admission control rate, sensor node should have a small slot of time to access to the channel and transmit data, employment of time division multiplexing (TDM) technique
	Link layer flooding	Attacking MAC	Drain battery power or DoS execution	Improving port security and intrusion detection systems
	Link layer jamming	Obtaining the packet arrival time probability distribution	Data packets are jammed	Comparing the interarrival time probability distribution for all sorts of packets.
	Spoofing and ARP-spoofing	Malicious nodes	Divert legitimate node traffic to the attacker	Secure cluster formation, authentication and anti-replay protection
	Unfairness	Executing exhaustion attack. Exploiting cooperative MAC protocols periodically.	Frequent intermittent connectivity which results in message delivery latency	Usage of small frames

it to another malicious node in the remote part of the network. The packet is replayed by the second node with malicious nodes that can receive these packets, which are deceived into believing that the legitimate node that sends the packets to the first malicious node is their single-hub neighbor and that the packets are from it. Any packet that is late while following the normal route to its destination is discarded.

c. **Node replication (Clone)**: The attacker's intension is to cause irregularities in the network by duplicating a compromised node and placing the malicious nodes in the different parts of the network. Once this is successful, the attacker can control how the network behaves by using few of the malicious nodes placed in the network (Conti et al., 2014.). Data aggregation, detection of anomaly, and voting protocols can be corrupted by inserting false data or subduing data that is legitimate.

d. **Routing attacks**:

In routing attacks, there are:

- **Misdirection**: The transmitted messages are deliberately redirected to the incorrect path by falsifying routing advertisement and updating this false information in the routing table of the neighboring nodes (Parno et al., 2005). The nodes under attack are erased completely after advertising the false routing information; hence, they cannot receive packets again.

- **Network partitioning**: The attacker implements partitioning in a fully connected network and the nodes in different partitions cannot communicate with one another even when they are still connected together in the network (Butun et al, 2020).

- **Routing loop**: A routing loop developed by spoofing routing updates is placed in a route path. Messages sent will continuously be routed through this same path. This event in due course will cause exhaustion of energy and finally failure in the network (Huang and Lee, 2003).

- **Spoofed, altered, or replay routing information**: Modification can be performed on routing information interchanged between nodes in the network. This can cause a harmful effect on the routing scheme (Butun et al, 2020).

e. **Sybil attack**: In sybil attacks, malicious nodes present different identities in the network. The attacker sends a conflicting routing path to the legitimate nodes, thus causing disorder. The effectiveness of fault-tolerance schemes in the network is depreciated and there is a remarkable threat to geographic routing protocols (Gupta et al, 2014). The summary of network layer threats, mode of attack, result of attack, and their countermeasure is shown in Table 7.3.

4. **Transport Layer Attack**

The connections between two or more nodes are controlled by the transport layer. Protocols that establish connections between these nodes are targeted (Raymond and Midkiff, 2008). Attacks in transport layer are:

a. **Desynchronization**: The original link established between two nodes is terminated by de-synchronizing their transmission. A typical way of implementing this attack is by transmitting sequences with wrong flag constantly to the two nodes that are communicating so that they can lose synchronization (Wood, and Stankovic, 2002).

TABLE 7.3

Summary of Network Layer Threats in Cyber-Physical System

OSI Layer	Security Threat	Mode of Attack	Result of Attack	Countermeasure
	Hello flooding	Gains longer transmission range than legitimate nodes	Disrupts nodes from receiving packets	Pair-wise authentication. Geographic routing
	Blackhole/ selfishness	Fake optimum route message	Packet dropping attack where all the relayed packets are dropped.	Anomaly detection-based intrusion detection systems
	Sinkhole	Attacker announces a false optimal path	Unfairness	Distributed monitoring methodology and geo-statistical sampling methodology for the verification of the bidirectional reliability is carried out by routing protocols
Network layer	Selective forwarding (grayhole)	Malicious node	Corrupts routing protocol and allocates many routes to itself	Multipath routing with random selection of paths to destination, use monitor nodes
	Wormhole	Malicious node	A passive eavesdrops of data, false topology creation or to be authenticated	Directional antenna, statistical analysis of multipath, geographic and temporal packet leaches
	Node replication (cloning)	Compromised node	Subvert data aggregation, the behavior of a network is diverted using elective protocols	Asymmetric key management schemes that are isolated.
	Routing attacks	Malicious node	Exhaustion of energy and failure in network	Routing protocols based on tree-path or implementing hop count limits for forwarded packets, formation of clusters that are secure, device authentication
	Sybil	Multiple identities of malicious node	This attack is aimed at exhausting the neighboring nodes computational and storage resources	Authentication

b. **The message queue telemetry transport (MQTT) exploit**: The MQTT is designed to enable communication between resource-constrained devices and operates using publish-and-subscribe messaging technique. Nevertheless, MQTT is not designed with security layer and as such the user is tasked with duty of addressing security problems (Singh, Rajan, Shivraj and Balamuralidhar, 2015).

c. **Session hijacking**: This is the process of exploiting and tampering a real communication session in order to have illegal access to a system's information or services. Because it is an IP network extension, session hijacking will be a big problem for internet of things (IoT) networks (Nikiforakis, Meert, Younan, Johns and Joosen, (2011).

d. **SYN-flooding**: The goal of an attacker is to deplete the node's memory or energy by sending a lot of false messages to it. A typical example is when an attacker sends several requests for connection without completing any of the request. This will overburden the buffer and the node will be dead in due course (Wood, and Stankovic, 2002). The summary of transport layer threats and their mode of attack, result of attack, and the countermeasure is given in Table 7.4.

5. **Application Layer Attack**

DOS attacks are used to attack the application layer. Obstruction or evasion can be executed on protocols such as fusion, data aggregation, node localization, association, and time synchronization. Attacks on application layers are:

a. **Constrained application protocol (CoAP) exploit**: COAP was designed as a replica of HTTP for resource-constrained IoT devices to enable them communicate with the Internet. The security of COAP has posed a lot of challenges (Rahman and Shah, 2016).

b. **False data injection**: This attack occurs in the semantic level of a network. The aim of the attacker is to alter the general measurement or reading outcome

TABLE 7.4

Summary of Transport Layer Threats in Cyber-Physical System

OSI Layer	Security Threat	Mode of Attack	Result of Attack	Countermeasure
Transport layer	Desynchronization	Disrupt communication established between two legitimate nodes	It is a resource exhaustion attack.	Device authorization
	MQTT exploit	Disrupt communication	Disrupt communication between nodes	Management of session keys and the issuance of SSL/TLS certificates or lightweight cryptographic mechanisms can be used.
	Session hijacking	Spoofs victim node IP address	DoS attacks	Software updating, end point security, having a biometric authentication for each user
	SYN-flooding	Open TCP connection with victim's node	DoS attacks	The deployment of ubiquitous security mechanism such as IPS and network equipment must have up-to-date installations

TABLE 7.5

Summary of Application Layer Threat in Cyber-Physical System

OSI Layer	Security Threat	Mode of Attack	Result of Attack	Countermeasure
Application Layer	COAP Exploit	DoS	Block attack, request delay attack, response delay and mismatch attack, relay attack, request fragment rearrangement attack	Providing extra layer protection using datagram transport layer security (DTLS), TLS, or OSCORE
	False Data Injection	Malicious nodes	Influence overall system measurement or reading	Implementing a statistical en-route filtering (SEF)
	Path-based DoS	Flooding an end-to-end communication route	Source to destination nodes on the path are affected	Authentication and anti-replay protection
	Reprogramming	False messages are sent to the nodes	Pushing them into unstable or dead state	Authentication and anti-replay protection. Authentication streams
	Sensor overwhelming	Sending false interference	The CPS devices are put in an unstable state	Sensor tuning, data aggregation

by using malicious nodes to inject false data, thus having a logical effect (Manandhar et al., 2014).

c. **Path-based DoS**: This is a DOS attack in the physical layer which follows the same procedure of overwhelming the end-to-end communication path of nodes with replayed or modified packets (Deng et al., 2005).

d. **Reprogramming**: When networks are re-programmed or patched, attackers may sniff into the network during this vulnerable time by injecting spurious data into the nodes and causing them to be unstable or dead (Ghildiyal et al., 2014).

e. **Sensor overwhelming**: The sensitivity of the sensor's measurement may be attacked or altered by sending false interference and overpowering these sensors with stimulations that are false. Table 7.5 shows the threats in application layer, mode of attack, result of attack, and their countermeasures.

Cyber-Physical Threat

1. Industrial Control Systems (ICS)

- The wireless potential of ICS can be utilized by the criminal attackers to control the ICS application and perhaps interrupt its functioning (Humayed et al., 2017).

- Customers with skills may alter the equipment physically or insert false information to misdirect the utility, thereby reducing the customers utility bill and causing financial loss to the company (Turk, 2005).

- The control center can be misinformed by the attacker. The physical threat could come by the attacker spoofing a temperature sensor by applying cold or heat to the sensor (Humayed et al., 2017).

- During war, a hostile nation may remotely target the critical infrastructure of its enemy nation by injecting malware or gaining access and controlling these devices in the field leading to shutting down, wrecking it or using it to generate pollution in the environment (Tsang, 2010).

2. Smart Grids

- The utility company billing system can be deceived by their customer by tampering the smart devices to reduce their electricity bill (Mo et al., 2011).
- Consumer's private information from the communication between the utility company and the smart meter may be hacked by armed robbers and used to accomplish their robbery attack (Sridhar et al, 2011).
- Smart grid infrastructure can be accessed remotely by a hostile nation to cause a blackout in a country (McDaniel, and McLaughlin, 2009).

3. Medical Devices

- A patient can be harmed by a criminal by simply gaining remote access to the patient's medical device and injecting false data with the aim of changing the state of device to cause undesired health condition to the patient. Furthermore, the device"s signal may be jammed to prevent communication between the devices, causing the device to be unstable and thus failing to deliver expected therapies to the patient (Halperin et al., 2008).
- The privacy and confidentiality of a patient may be invaded by an attacker by intercepting the communication between the medical device and the hospital or having an unauthorized access to the database of the hospital with the goal of exposing the patient's sickness or altering the patient's data which could lead to wrong drug prescription and eventually death (Le et al., 2011).
- In cyberwar, country's political figures can be assassinated remotely by the hostile nation by taking advantage of their medical device wireless communication and inflicting serious health condition or death.

4. Smart Cars

- The privacy invasion of a car owner may be invaded by an attacker by intercepting the private communication of the car via TCU vulnerabilities (Checkoway et al., 2011). Smart cars may be caused to lose control or collide by taking advantage of the wireless interface weaknesses and attacking the car's ECUs.
- GPS navigation system used to direct drivers can be exploited by law enforcement to track the car (Brooks et al., 2008).
- Car manufacturers can deliberately sell the cars' log data gathered in the ECU without consent from the car owners. This leads to privacy invasion (Hoppe et al., 2011).
- Cyberwar can be issued against a country by a hostile nation by targeting the transportation sector of the country and causing collisions among smart cars on the road (Checkoway et al., 2011).

5. Smart House

The smart house consists of sensors and actuators that are wirelessly interconnected. Any compromise or alteration to these sensitive devices can lead to a major risk in the network. The smart house can be attacked in two ways:

- Pricing cyberattack

  The smart house can be attacked by abnormally increasing the power consumption of the smart house by altering the pricing schedules of electricity so as to increase the voltage and current of the smart house (Wurm et al, 2016).

- Privacy invasion: Hackers may intercept communications within the devices in the smart house leading to privacy invasion. These sensors and actuators may be injected with false data with the aim of changing the state of these devices to cause undesired outcome to the smart house residents which can lead to death.

- Energy theft: The flow of energy in a smart house can be manipulated to reduce the load of the attacker's house and increase the load of the smart house, thereby increasing the energy bill of the house and decreasing the attacker's bill. This is because bills in a particular community is calculated based on each energy consumption of a smart house (Qiu et al., 2017).

Causes of Vulnerabilities in CPS

The causes of vulnerabilities in CPS are:

1. **Increased connectivity**: The connectivity in CPS is increasing due to the increment in services offered by manufacturers. These services depend on technologies based on wireless connectivity and open networks. Attacks on ICS usually were internal, but of recent, attacks are now from external sources (internet-based) (Byres and Lowe, 2004). Furthermore, majority of these devices used in the field are directly connected to the internet for quick reaction to an event, for simple configuration and management, result in increase in attack (Leverett and Wightman, 2013).

2. **Heterogeneity**: CPS components like proprietary components, commercial off-the-shelf (COTS), and third party used in designing CPS applications are heterogeneous in nature. They are most likely multivendor systems with each having security vulnerabilities (Ericsson, 2010). Furthermore, the integrated heterogeneous components used are concealed, which can lead to unforeseen performances. These pave way to fundamental susceptibility (Amin et al., 2013).

3. **Isolation assumption**: The preliminary design of CPS was "security by Obscurity". The main idea was to design CPS that is dependable and secured by protecting these systems from the outside world. Traditional ICS and power grid security depended on the securing these systems by protecting them from the influence of the outside world, thereby performing operations locally (Halperin et al., 2008). Currently, CPS applications developed are not isolation-based in design but instead, they are built with multi-connectivity leading to increase in attack.

7.8 Mitigation Schemes and Control Mechanisms for CPS Attack

A. Cryptographic Solutions

Cryptographic measures are primarily used for the protection of end-to-end communication channels from active and passive CPS threats and attacks, in addition to unauthorized access and interception, particularly in SCADA systems (American Gas Association,

2005). It is very difficult to implement cryptographic protocols on CPS systems due to their resource-constrained nature. As a result, the focus should not just be on data security but also on maintaining and guaranteeing the efficiency of the overall system process. In Kocabas (2016), the authors conducted their own survey on traditional and emerging encryption systems that may be used to for secure storage of CPS data. They evaluated and described popular encryption and authentication mechanisms (Kirkpatrick, 2009) for securing distributed energy resources (DER) systems. Ding et al. (2018) offered an overview of current improvements in industrial CPS security control and attack detection, particularly against masquerade, replay, and flooding attacks. Sklavos et al. (2016) looked at the implementation efficiency of CPS security requirements. They pointed out that implementing user authentication can protect individual CPS node from malicious activities. A novel approach was developed for understanding security-related risk and threats (Hahn et al., 2015). The proposed approach carried out a thorough examination of CPS attack factors such as the adversary and its objectives, cyber exploitation, control mechanisms, and specific features of CPS. A complete ICS security guideline was provided in Stouffer et al. (2007) which relates to operational controls such as pervasive security mechanisms and traditional cryptographic solutions. In Nicholson (2012), a simulation attack was performed using social engineering, and phishing techniques security specialists used phishing to gain access to the data of employees as a result of their poor cybersecurity hygiene. A robust security framework for multilevel security evaluation was proposed by Sharma et al. (2018). As a result, they offered a comprehensive assessment of whether the proposed security framework was suitable for resource-constrained network applications. The proposed security framework provided recommendations for administrators of enterprise networks during initial deployment in order to accomplish the appropriate security demands. As a result, this chapter categorizes various solutions based on the NIST Cybersecurity requirements which are discussed below.

Confidentiality: It is critical to secure CPS communication links and as such, numerous cryptographic solutions were provided. Zhang et al. (2014) proposed an encryption technique that uses a compression algorithm. This algorithm was a symmetric lightweight authentication algorithm. Similarly, a lightweight encryption mechanism was proposed by Bogdanov et al. (2007) and a block cipher with very little delay for ubiquitous computing applications. Block cipher was used due to its low cost and latency, as well as its ability to offer cryptographic blocks for CPS devices with low communication and computational capabilities. Shahzad et al. (2015) proposed an end-to-end security framework for Modbus communication network. The system was developed to prevent confidentiality threats such as traffic analysis and eavesdropping. This resulted in additional overhead in the conversion of the plaintexts to ciphertexts. In Industrial IoT applications, a "bump-in-the-wire" approach was proposed by the American Gas Association (AGA) for encryption of CPS systems; however, the proposed solution had very high computational delay (Rubio-Hernán, 2016). A security framework was proposed by Vegh et al. (2016). The proposed system was based on a hierarchical cryptography which was developed from generated ElGamal algorithm for the protection of CPS networks. Zhou et al. (2020) proposed a lightweight key management scheme for event-driven CPS applications, including vehicular ad-hoc networks (VANETs) (Zeadally et al., 2012). The authors claimed that their proposed solution was secure and dependable for critical applications. He et al. (2018) introduced an attribute-based encryption scheme for CPS applications that depended on cloud infrastructure. Their results show that the proposed scheme is lightweight and secure and has the recommended security requirements, including accountability. A blockchain-based architecture was proposed for industrial CPS by Zhao et al. (2018). The proposed scheme met privacy requirements by providing hybrid encryption and data secrecy. Sepúlveda et al. (2019) leveraged on the traditional security mechanism (asymmetric authentication) for CPS frameworks to provide security for heterogeneous CPS applications. Their proposed approach was based on datagram transport layer security (DTLS) using post quantum computing.

Integrity: Keeping CPS devices in good working order necessitates the prevention of logical or physical compromise of real-time packets from source, relay, and destination nodes. As a result, various solutions are provided. To mitigate software reconfiguration and attacks on network on CPS, Harshe et al. (2015) proposed a secure trust management (TAIGA). framework for CPS networks. TAIGA ensures that industrial and supervisory nodes in control plants are secure and also integrates a reputation-based privacy preserving backup controller. A shadow security unit "SSU" was presented for low power device security. This was used with remote terminals for secure communication with SCADA systems (Luallen, 2013). This system was implemented as an additional security to the SIEM systems. Another proposed approach was described in Ghaleb et al. (2018) to prevent man-in-the-middle, replay, and command modification attacks by providing an encryption level for transported packets and using hardware cipher models. Cao et al. (2013) developed a layered strategy to securing sensitive data. Their methods relied on a hash system. As a result, data interception was prevented using a layered hierarchical security policy.

Availability: It is critical to keep CPS devices operational. As a result, many strategies to reduce and solve availability concerns were proposed in the literature. Amin et al. (2011) proposed a process control system for integrity test. Their simulation results show that the system was robust against integrity and DoS attacks with considerable computational overhead. Cárdenas et al. (2011) proposed the prioritization of cryptographic schemes to provide layered security. They showed that their system was capable of overcoming DoS attacks. Gao et al. (2018) developed a testing solution for SCADA network solution. This scheme was developed for network emulations and process control interactions between controllers in the CPS framework. Alves et al. (2018) proposed an IPS architecture based on machine learning which used an open-source programmable logic controller to prevent the activities of malicious users. The proposed framework was able to defend the system from DoS, masquerading, wormhole, and traffic analysis attacks.

B. Non-cryptographic Solutions

Several ubiquitous security mechanisms which are non-cryptographic approaches have been proposed for CPS networks. Majority of these security mechanisms deploy intrusion detection systems (IDS), firewalls, and honeypots. As a result, several strategies proposed in literature are briefly discussed.

Intrusion Detection Systems

The implementation of IDS in CPS networks is as a result of the various network configurations available. When it comes to detection, setup, pricing, and network location, each IDS methodology has its own set of benefits and downsides. Several cybersecurity strategies were deployed to mitigate the threats against CPS networks, according to Almohri et al. (2013). Shu et al. (2015) and Xu et al. (2016) clearly point out that a cybersecurity threat can be recognized when a proper a cybersecurity framework is developed and modeled from spatiotemporal data. They clearly pointed out that pervasive security mechanism deployed in literature deal with specific threats against specific applications, such as unmanned aerial vehicles (UAVs) (Mitchell and Ray, 2013), industrial control processes (Urbina, 2016), and smart grids (Sridhar, 2012). Zimmer et al. (2010) leveraged the possibility of a worst-case execution time by employing static application analysis to acquire information in order to detect wormhole attacks in the form of code injection attacks. Another IPS strategy is the behavior-rule specification-based technique for using IDS in medical CPS proposed in Mitchell and Ray (2014). The authors also demonstrated how behavior rules may be transformed into a control operation that can detect anomaly from the given specification of medical devices.

Intrusion Detection Schemes

Signature-based, anomaly-based, behavior-based, and hybrid-based intrusion detection approaches are the four basic IDS methodologies. In reality, these methodologies were provided in Zarpelão et al. (2017), whereas testing methods and procedures were divided into five primary categories based on their detection mechanism.

a. **Anomaly-based IDS**: Compares the system's operations in real time and can trigger an alert anytime a divergence from typical behavior is discovered. This type of approach, however, has a large false-positive rate (Raza et al., 2013). Hong et al. (2009) proposed a botnet identification strategy which leveraged on the anomaly-based detection strategy by the computation of summation value for profiling based on the proposed metrics. This was accomplished prior to the system monitoring network traffic and raising an alert if a measure deviates from the previously defined computed averages. Gupta et al. (2013) proposed their own architecture for a wireless IDS, using the requisite computational intelligence methods to build typical profile behavior. Furthermore, for each IP address assigned, a unique normal behavior profile will be established. Lee et al. (2014) proposed that the consumption of energy be classed as a parameter and utilized to analyze the behavior of each node. For each mesh-under routing scheme and route-over routing scheme, a regular energy consumption model was defined, with each node monitoring its own energy utilization. If the node does not act accordingly, the IDS flags it as an anomaly and the system gets rid of it from the network.

   Summerville et al. (2015) used a bit-pattern matching strategy that performs a feature selection to produce an anomaly detection approach based on deep packet inspection, which is focused at reducing the run-on CPS devices with restrictions. Four attacks were used to evaluate the robustness of the proposed scheme and the results showed low false-positive rates. Thanigaivelan et al. (2016) successfully designed an IoT-distributed internal anomaly detection system that checks the data rate and packet size of the node. Furthermore, in Pongle and Chavan (2015), an IDS framework is proposed for the identification of wormhole attacks in IoT devices, as well as three major techniques for detecting anomalies in network. The proposed scheme was able to identify and remove malicious nodes from the network. Demertzis et al. (2017) proposed an IDS mechanism-based neural network. They used structural and temporal data in the characterization of network activities. The proposed approach had the ability to identify irregularities or deviations in behavior that are linked with APT attacks. According to the authors, SOCCADF is well suited to complex problems and applications involving large amounts of data. The authors concluded that SOCCADF outperforms the other techniques with lower computational overheads and high efficiency and reliability.

Signature-based detection: This IDS strategy is one of the simplest forms of IDS methods to deploy. It is, however, only useful for recognizing known threats. Signature-based IDS is susceptible to threats that are unknown. Signature-based IDS has been described as very effective with high precision even with its limitations. The signatures are frequently updated because their patches are unknown, which makes the detection of new and old attacks very difficult (Liao et al., 2013; Vacca, 2012). Oh et al. (2014) compared the payload of the transmitted data and the attack signatures with the intention of reducing computational cost. Liu et al. (2011) proposed an "artificial immune system" (AIS) strategy developed with immune cells that distinguish and classify malicious and non-malicious entities by signature matching. Kasinathan et al. (2013) paid attention to DoS attack detection in resource-constrained CPS networks. They developed a signature-based IDA by the adoption of "Suricata4" for use on 6LoWPAN networks. This was carried to lower the false alarm rate in the network

b. **Behavior-based**: The term "behavior-based" refers to polices and boundaries used to specify intended action of network elements such as workstation and services. Intrusion can be detected once there is a deviation in the behavior of the network. This method greatly relies on structural and spatiotemporal modeling of data. Behavior-based detection works similarly to anomaly-based detection, with the exception that specification-based systems require a human expert to design each specification rule explicitly. As a result, the false-positive rate is lower than with anomaly-based detection (Mitchell and Ing-Ray, 2014; Butun et al., 2013). As a result, there will be no need for training because they are installed and ready to use right away. However, this method is not appropriate for all cases, and it may have very high delay. Because of interoperability issues, most CPS frameworks deploy middleware architectures which must be protected from DoS attacks as proposed by Misra et al. (2011). The authors develop a security system using a threshold mechanism so that the system detects when the transmission request exceeds the specified threshold. Le et al. (2011) proposed a new specification-based technique that intended to detect RPL attacks by specifying RPL behavior through network monitoring operations and malicious action detection. According to Zarpelão et al. (2017), their experimentation resulted in a high true-positive rate with low false-positive rates throughout, while also incurring considerable energy cost when compared with a normal RPL network. Amaral et al. (2014) developed a specification-based IDS that allows the network administrator to create and maintain rules to identify any potential attack. When a rule is broken, the IDS immediately sends an alarm to the event management system (EMS), which correlates the alerts for all available nodes in a network. The network administrator's expertise, as well as his experience and abilities combined, was critical to the success of Misra et al.'s (2011) and Amaral et al.'s (2014) approaches. As a result, any incorrect specifications will result in an extremely high false-positive rate and/or a high false-negative rate, posing a potentially major risk to the network's security.

c. **Hybrid IDS**: It is built on the use of signature-based and anomaly-based detection approaches that are specification based in order to maximize their benefits while limiting their downsides. Raza et al. (2013) introduced SVELTE, a hybrid IDS that strikes the correct balance between storage cost of signature-based approaches and computing cost of anomaly-based methods. Krimmling et al. (2014) used the IDS evaluation framework they introduced to test their anomaly- and signature-based

IDS. Their findings demonstrated that each strategy failed to detect specific threats on its own. As a result, the authors merged various methods in order to cover and identify a wider spectrum of attacks. Cervantes et al. (2015) published the Intrusion Detection of Sinkhole Assaults on 6LoWPAN for Internet of Things (INTI), which combines an anomaly-based technique with a packet exchange between these nodes to detect and isolate sinkhole attacks. This was accomplished by extracting the evaluation node based on both trust and reputation utilizing the specification-based method. Cervantes et al. constructed a situation in which INTI IDS detected sinkholes at a rate of up to 92% when comparing SVELTE 9 (Pongle and Gurunath, 2015) to INTI IDS. The rate has only reached 75% in the case of a set situation. In any case, compared to SVELTE, it has a low rate of false-positives and false-negatives.

7.9 Challenges and Open Issues

CPS have a tremendous capacity to develop new markets and offer solutions to societal problems, but they impose stringent quality, safety, security, and privacy requirements. Fundamental scientific research is required to attain a predictable degree of verification and measurement quality in order to combat external and internal changes successfully. Future study directions include the following tasks, based on the aforementioned analysis of the most recent CPS security studies:

a. An examination of the major CPS issues that have arisen as a result of the rapid development of cyber and physical threats reveals the need for the development of a dependable and fault-tolerant architecture that assures a high level of security and cost-effectiveness.

b. Process noises and uncertainties in system model parameters are not taken into account in most existing studies. They also presume that the attacked system's state trajectory is fully specified and can be precisely measured. These assumptions in previous studies need to be revisited. A problem worth pursuing is how to model the attack under more realistic assumptions. It is critical to investigate the attackers' intents and behaviors in order to create an effective defense strategy. Knowing how various types of attacks affect system performance can provide theoretical guidance for attack detection and resilient control.

c. With the growth of CPS security approaches, a testbed is needed to evaluate developing ideas, methods, and strategies. The testbed should provide a realistic and real-time environment in which to conduct attack-defense experiments. In general, CPS security is a difficult subject that necessitates techniques from a variety of domains, including hybrid systems, discrete event systems, networked control systems, and big data analysis.

d. Many of the aforementioned authentication strategies are not particularly suited for a secure appliance because of the lack of multifactor authentication procedures to safeguard CPS systems from unwanted users and access.

e. Regardless of the fact that numerous IDS types, such as anomaly-based, behavior-based, and signature-based are available, they are typically employed in IoT-based domains and are not specifically designed to protect CPS systems.

7.10 Conclusion

This chapter provided a summary of the state-of-the-art security threats in CPS, as well as contemporary ways to tackle the ever-increasing risks. Because of the fast adoption of technologies such as IoT, smart home/automotive/energy, and a general deeper link between the cyber and physical world, the problem will become increasingly relevant. A brief assessment of CPS threats was presented, as well as proposals for CPS security requirements. There is also a taxonomy of attack types and security mechanisms to mitigate these attacks. Finally, we hope this article will be useful to both industry and academia in CPS security.

References

Aazam, M., S. Zeadally, and K. Harras "Deploying Fog Computing in Industrial Internet of Things and Industry 4.0." *IEEE Transactions on Industrial Informatics* 14, (2018): 4674–4682.

Amaral, João P., Luís M. Oliveira, Joel J.P.C. Rodrigues, Guangjie Han, and Lei Shu "Policy and network-based intrusion detection system for IPv6-enabled wireless sensor networks." In *2014 IEEE International Conference on Communications (ICC)*, pp. 1796–1801. IEEE, 2014.

Almohri, Hussain MJ, Danfeng Yao, and Dennis Kafura "Process authentication for high system assurance." *IEEE Transactions on Dependable and Secure Computing* 11, no. 2 (2013): 168–180.

Alves, Thiago, Rishabh Das, and Thomas Morris. "Embedding encryption and machine learning intrusion prevention systems on programmable logic controllers." *IEEE Embedded Systems Letters* 10, no. 3 (2018): 99–102.

American Gas Association. *Cryptographic protection of SCADA communications part 1: Background, policies and test plan*. No. 12 Part 1. AGA Report, 2005.

Amin, Saurabh, Galina A. Schwartz, and Alefiya Hussain "In quest of benchmarking security risks to cyber-physical systems." *IEEE Network* 27, no. 1 (2013): 19–24.

Amin, Saurabh, Galina A. Schwartz, and S. Shankar Sastry "On the interdependence of reliability and security in networked control systems." In *2011 50th IEEE Conference on Decision and Control and European Control Conference*, pp. 4078–4083. IEEE, 2011.

Ashibani, Y. and Mahmoud, Q. Cyber-physical systems security: Analysis, challenges and solutions. *Computers Security* 68 (2017): 81–97. https://www.sciencedirect.com/science/article/pii/S0167404817300809.

Bogdanov, Andrey, Lars R. Knudsen, Gregor Leander, Christof Paar, Axel Poschmann, and Matthew J.B. Robshaw, Yannick Seurin, and Charlotte Vikkelsoe. "PRESENT: An ultra-lightweight block cipher." In *International Workshop on Cryptographic Hardware and Embedded Systems*, pp. 450–466. Springer, Berlin, Heidelberg, 2007.

Borgohain, Tuhin, Uday Kumar, and Sugata Sanyal "Survey of security and privacy issues of internet of things." *arXiv preprint arXiv:1501.02211* (2015).

Brar, Suman and Mohit Angurala "Review on grey-hole attack detection and prevention." *International Journal of Advance research, Ideas and Innovations in Technology* 2, no. 5 (2016): 1–4.

Brooks R.R., S. Sander, J. Deng, and J. Taiber "Automotive system security: challenges and state-of-the-art." In *Proceedings of the 4th Annual Workshop on Cyber Security and Information Intelligence Research: Developing Strategies to Meet the Cyber Security and Information Intelligence Challenges Ahead*, pp. 1–3. 2008.

Butun, Ismail, Patrik Österberg, and Houbing Song "Security of the Internet of Things: Vulnerabilities, attacks, and countermeasures." *IEEE Communications Surveys & Tutorials* 22, no. 1 (2019): 616–644.

Butun, Ismail "Prevention and detection of intrusions in wireless sensor networks", 2013.

Butun, Ismail, Salvatore D. Morgera, and Ravi Sankar "A survey of intrusion detection systems in wireless sensor networks." *IEEE Communications Surveys & Tutorials* 16, no. 1 (2013): 266–282.

Byres, Eric and Justin Lowe "The myths and facts behind cyber security risks for industrial control systems." *Proceedings of the VDE Kongress* 116, (2004): 213–218.

Cao, Huayang, Peidong Zhu, Xicheng Lu, and Andrei Gurtov "A layered encryption mechanism for networked critical infrastructures." *IEEE Network* 27, no. 1 (2013): 12–18.

Cárdenas, Alvaro A., Saurabh Amin, Zong-Syun Lin, Yu-Lun Huang, Chi-Yen Huang, and Shankar Sastry "Attacks against process control systems: risk assessment, detection, and response." In *Proceedings of the 6th ACM Symposium on Information, Computer and Communications Security*, pp. 355–366. 2011.

Cervantes, Christian, Diego Poplade, Michele Nogueira, and Aldri Santos "Detection of sinkhole attacks for supporting secure routing on 6LoWPAN for Internet of Things." In *2015 IFIP/IEEE International Symposium on Integrated Network Management (IM)*, pp. 606–611. IEEE, 2015.

Chawla, Astha, Animesh Singh, Prakhar Agrawal, Bijaya Ketan Panigrahi, Bhavesh R. Bhalja, and Kolin Paul "Denial-of-Service attacks pre-emptive and detection framework for synchrophasor based wide area protection applications." *IEEE Systems Journal* 16, (2021): 1570–1581.

Checkoway, Stephen, Damon McCoy, Brian Kantor, Danny Anderson, Hovav Shacham, Stefan Savage, Karl Koscher, Alexei Czeskis, Franziska Roesner, and Tadayoshi Kohno "Comprehensive experimental analyses of automotive attack surfaces." In *USENIX Security Symposium*, vol. 4, (2011): 447–462, p. 2021.

Chun, Ingeol, Jeongmin Park, Wontae Kim, Woochun Kang, Haeyoung Lee, and Seungmin Park "Autonomic computing technologies for cyber-physical systems." In *2010 The 12th International Conference on Advanced Communication Technology (ICACT)*, vol. 2, pp. 1009–1014. IEEE, 2010.

Conti, Mauro., Roberto Di Pietro, and Angelo Spognardi "Clone wars: Distributed detection of clone attacks in mobile WSNs." *Journal of Computer and System Sciences* 80, no. 3 (2014): 654–669.

Demertzis, Konstantinos., Lazaros Iliadis, and Stefanos Spartalis "A spiking one-class anomaly detection framework for cyber-security on industrial control systems." In *International Conference on Engineering Applications of Neural Networks*, pp. 122–134. Springer, Cham, 2017.

Deng, J., Han, R., and Mishra, S. "Defending against path-based dos attacks in wireless sensor networks," In *Proceedings of the 3rd ACM workshop on Security of ad hoc and sensor networks*, pp. 89–96. ACM, 2005.

Ding, Derui, Qing-Long Han, Yang Xiang, Xiaohua Ge, and Xian-Ming Zhang "A survey on security control and attack detection for industrial cyber-physical systems." *Neurocomputing* 275 (2018): 1674–1683.

Ericsson, Göran N. "Cyber security and power system communication—essential parts of a smart grid infrastructure." *IEEE Transactions on Power Delivery* 25, no. 3 (2010): 1501–1507.

Di Antonio, Ferdinando, Paul Ezhilchelvan, Michael Dales, and Jon Crowcroft "A qos-negotiable middleware system for reliably multicasting messages of arbitrary size." In *Ninth IEEE International Symposium on Object and Component-Oriented Real-Time Distributed Computing (ISORC'06)*, p. 8. IEEE, 2006.

Gao, Haihui, Yong Peng, Kebin Jia, Zhonghua Dai, and Ting Wang "The design of ics testbed based on emulation, physical, and simulation (eps-ics testbed)." In *2013 Ninth International Conference on Intelligent Information Hiding and Multimedia Signal Processing*, pp. 420–423. IEEE, 2013.

Ghaleb, Asem, Sami Zhioua, and Ahmad Almulhem "On PLC network security." *International Journal of Critical Infrastructure Protection* 22 (2018): 62–69.

Ghildiyal, Sunil, Amit Kumar Mishra, Ashish Gupta, and Neha Garg "Analysis of denial of service (dos) attacks in wireless sensor networks." *IJRET: International Journal of Research in Engineering and Technology* 3 (2014), ISSN 2319-1163.

Gries, Stefan, Marc Hesenius, and Volker Gruhn "Cascading data corruption: About dependencies in cyber-physical systems: Poster." In *Proceedings of the 11th ACM International Conference on Distributed and Event-based Systems*, pp. 345–346. 2017.

Gupta, Astha, Ramanuj Maurya, R.K. Roy, Samir V. Sawant, and Hemant Kumar Yadav "AFLP based genetic relationship and population structure analysis of Canna—An ornamental plant." *Scientia Horticulturae* 154 (2013): 1–7.

Gupta Hari, Prabhat, S. V. Rao, Amit Kumar Yadav, and Tanima Dutta. "Geographic routing in clustered wireless sensor networks among obstacles." *IEEE sensors Journal* 15, no. 5 (2014): 2984–2992.

Gupta, R., S. Tanwar, F. Al-Turjman, P. Italiya, A. Nauman, and S. Kim "Smart contract privacy protection using AI in cyber-physical systems: Tools, techniques and challenges." *IEEE Access* 8 (2020): 24746–24772.

Hadley, M. D., K. A. Huston, and T. W. Edgar "AGA-12, part 2 performance test results." *Pacific Northwest National Laboratories* (2007).

Hahn, Adam, Roshan K. Thomas, Ivan Lozano, and Alvaro Cardenas "A multi-layered and kill-chain based security analysis framework for cyber-physical systems." *International Journal of Critical Infrastructure Protection* 11 (2015): 39–50.

Halperin, Daniel., Thomas S. Heydt-Benjamin, Kevin Fu, Tadayoshi Kohno, and William H. Maisel. "Security and privacy for implantable medical devices." *IEEE Pervasive Computing* 7, no. 1 (2008): 30–39.

Hamdan, M., M. Mahmoud, and U. Baroudi "Event-triggering control scheme for discrete time cyberphysical Systems in the presence of simultaneous hybrid stochastic attacks." *ISA Transactions* 122 (2021): 1–12. https://www.sciencedirect.com/science/article/pii/S001905782100224X.

Harshe, Omkar A., N. Teja Chiluvuri, Cameron D. Patterson, and William T. Baumann "Design and implementation of a security framework for industrial control systems." In *2015 International Conference on Industrial Instrumentation and Control (ICIC)*, pp. 127–132. IEEE, 2015.

He, Qian, Ning Zhang, Yongzhuang Wei, and Yan Zhang "Lightweight attribute based encryption scheme for mobile cloud assisted cyber-physical systems." *Computer Networks* 140 (2018): 163–173.

Heng, Stefan "Industry 4.0: Huge potential for value creation waiting to be tapped." *Deutsche Bank Research* (2014): 8–10.

Hoppe, T., S. Kiltz, and J. Dittmann Security threats to automotive can networks—Practical examples and selected short-term countermeasures. In *Proceedings of the 27th International Conference on Computer Safety, Reliability, and Security, SAFECOMP '08*, pp. 235–248. Springer-Verlag, Berlin, Heidelberg, 2011.

Howladar, P., P. Roy, and H. Rahaman "MEDA based biochips: Detection, prevention and rectification techniques for cyberphysical attacks." *IEEE/ACM Transactions on Computational Biology and Bioinformatics* 19 (2021): 2345–2355.

Hu, Fei, and Neeraj K. Sharma "Security considerations in ad hoc sensor networks." *Ad Hoc Networks* 3, no. 1 (2005): 69–89.

Hu Y.-C., Adrian Perrig, and David B. Johnson "Packet leashes: A defense against wormhole attacks in wireless networks." In *IEEE INFOCOM 2003. Twenty-second Annual Joint Conference of the IEEE Computer and Communications Societies (IEEE Cat. No. 03CH37428)*, vol. 3, pp. 1976–1986. IEEE, 2003.

Huang, Y. A., and W. Lee "A cooperative intrusion detection system for ad hoc networks," in *Proceedings of the 1st ACM Workshop on Security of Ad Hoc and Sensor Networks*, pp. 135–147. ACM, 2003

Humayed, Abdulmalik, Jingqiang Lin, Fengjun Li, and Bo Luo "Cyber-physical systems security—A survey." *IEEE Internet of Things Journal* 4, no. 6 (2017): 1802–1831.

Islam, S., D. Kwak, M. Kabir, M. Hossain, and K. Kwak The Internet of Things for health care: A comprehensive survey. *IEEE Access* 3 (2015): 678–708.

Johnson, Robert E. "Survey of SCADA security challenges and potential attack vectors." In *2010 International Conference for Internet Technology and Secured Transactions*, pp. 1–5. IEEE, 2010.

Juma, M., and K. Shaalan (2020). Cyberphysical systems in the smart city: Challenges and future trends for strategic research. In Aboul Ella Hassanien, Ashraf Darwish (Eds.), *Swarm Intelligence for Resource Management in Internet of Things* (pp. 65–85). Academic Press. https://www.sciencedirect.com/science/article/pii/B9780128182871000085

Karlof, C., and D. Wagner "Secure routing in wireless sensor networks: Attacks and countermeasures," *Ad Hoc Networks* 1, no. 2 (2003): 293–315.

Kasinathan, Prabhakaran, Claudio Pastrone, and Maurizio A. Spirito, and Mark Vinkovits "Denial of-Service detection in 6LoWPAN based Internet of Things". In *5th International Conference on the Internet of Things (IOT)*, pp. 30–36. 2013.

Kaur, Rupinder, and Parminder Singh "Review of black hole and grey hole attack." *The International Journal of Multimedia & Its Applications* 6, no. 6 (2014): 35.

Khaitan, S., and J. McCalley "Design techniques and applications of cyberphysical systems: A survey." *IEEE Systems Journal* 9 (2015): 350–365.

Kibirige, George W., and Camilius Sanga. "A survey on detection of sinkhole attack in wireless sensor network." *arXiv preprint arXiv:1505.01941* (2015).

Kocabas, Ovunc, Tolga Soyata, and Mehmet K. Aktas "Emerging security mechanisms for medical cyber-physical systems." *IEEE/ACM Transactions on Computational Biology and Bioinformatics* 13, no. 3 (2016): 401–416.

Krimmling, Jana, and Steffen Peter "Integration and evaluation of intrusion detection for CoAP in smart city applications." In *2014 IEEE Conference on Communications and Network Security*, pp. 73–78. IEEE, 2014.

Kumar, J. Sathish, and Dhiren R. Patel "A survey on internet of things: Security and privacy issues." *International Journal of Computer Applications* 90, no. 11 (2014). DOI:10.5120/15764-4454

Le, Anhtuan., Jonathan Loo, Yuan Luo, and Aboubaker Lasebae "Specification-based IDS for securing RPL from topology attacks." In *2011 IFIP Wireless Days (WD)*, pp. 1–3. IEEE, 2011.

Lee, Jay, Behrad Bagheri, and Hung-An Kao. "A cyber-physical systems architecture for Industry 4.0-based manufacturing systems." *Manufacturing Letters* 3 (2015): 18–23.

Leverett, Éireann, and Reid Wightman. "Vulnerability inheritance in programmable logic controllers." In *Proceedings of the Second International Symposium on Research in Grey-Hat Hacking*, 2013.

Liao, Hung-Jen, Chun-Hung Richard Lin, Ying-Chih Lin, and Kuang-Yuan Tung "Intrusion detection system: A comprehensive review." *Journal of Network and Computer Applications* 36, no. 1 (2013): 16–24.

Liu, Y., Li, Y., and Man, H. "Mac layer anomaly detection in ad hoc networks." In *Information Assurance Workshop, 2005. IAW'05. Proceedings from the Sixth Annual IEEE SMC*, pp. 402–409. IEEE, 2005.

Luallen, Matthew E. "Sans scada and process control security survey." *A SANS Whitepaper, February* (2013).

Manandhar, Kebina, Xiaojun Cao, Fei Hu, and Yao Liu. "Detection of faults and attacks including false data injection attack in smart grid using Kalman filter." *IEEE Transactions on Control of Network Systems* 1, no. 4 (2014): 370–379.

Patrick, McDaniel, and Stephen McLaughlin "Security and privacy challenges in the smart grid." *IEEE Security & Privacy* 7, no. 3 (2009): 75–77.

Misra, Sudip, P. Venkata Krishna, Harshit Agarwal, Antriksh Saxena, and Mohammad S. Obaidat "A learning automata based solution for preventing distributed denial of service in internet of things." In *2011 International Conference on Internet of Things and 4th International Conference on Cyber, Physical and Social Computing*, pp. 114–122. IEEE, 2011.

Mitchell, Robert, and Ing-Ray Chen. "A survey of intrusion detection techniques for cyber-physical systems." *ACM Computing Surveys (CSUR)* 46, no. 4 (2014): 1–29.

Mitchell, Robert, and Ray Chen. "Adaptive intrusion detection of malicious unmanned air vehicles using behavior rule specifications." *IEEE Transactions on Systems, Man, and Cybernetics: Systems* 44, no. 5 (2013): 593–604.

Yilin, Mo, Tiffany Hyun-Jin Kim, Kenneth Brancik, Dona Dickinson, Heejo Lee, Adrian Perrig, and Bruno Sinopoli. "Cyber–physical security of a smart grid infrastructure." *Proceedings of the IEEE* 100, no. 1 (2011): 195–209.

Nicholson, Andrew, Stuart Webber, Shaun Dyer, Tanuja Patel, and Helge Janicke. "SCADA security in the light of Cyber-Warfare." *Computers & Security* 31, no. 4 (2012): 418–436.

Nikiforakis, Nick, Wannes Meert, Yves Younan, Martin Johns, and Wouter Joosen. "SessionShield: Lightweight protection against session hijacking." In *International Symposium on Engineering Secure Software and Systems*, pp. 87–100. Springer, Berlin, Heidelberg, 2011.

Bryan, Parno, Adrian Perrig, and Virgil Gligor "Distributed detection of node replication attacks in sensor networks." In *2005 IEEE Symposium on Security and Privacy (S&P'05)*, pp. 49–63. IEEE, 2005.

Pongle, Pavan, and Gurunath Chavan. "Real time intrusion and wormhole attack detection in internet of things." *International Journal of Computer Applications* 121, no. 9 (2015): 1–9.

Qiu, Tie, Kaiyu Zheng, Houbing Song, Min Han, and Burak Kantarci "A local-optimization emergency scheduling scheme with self-recovery for a smart grid." *IEEE Transactions on Industrial Informatics* 13, no. 6 (2017): 3195–3205.

Rad, Ciprian-Radu, Olimpiu Hancu, Ioana-Alexandra Takacs, and Gheorghe Olteanu "Smart monitoring of potato crop: a cyber-physical system architecture model in the field of precision agriculture." *Agriculture and Agricultural Science Procedia* 6 (2015): 73–79.

Rahman, R. A. and Shah, B. "Security analysis of iot protocols: A focus in coap," In *Big Data and Smart City (ICBDSC), 2016 3rd MEC International Conference on* pp. 1–7. IEEE, 2016.

Raymond, David R., and Scott F. Midkiff. "Denial-of-service in wireless sensor networks: Attacks and defenses." *IEEE Pervasive Computing* 7, no. 1 (2008): 74–81.

Raza, Shahid, Linus Wallgren, and Thiemo Voigt. "SVELTE: Real-time intrusion detection in the Internet of Things." *Ad Hoc Networks* 11, no. 8 (2013): 2661–2674.

Rubio-Hernán, José, Luca De Cicco, and Joaquin Garcia-Alfaro. "Revisiting a watermark-based detection scheme to handle cyber-physical attacks." In *2016 11th International Conference on Availability, Reliability and Security (ARES)*, pp. 21–28. IEEE, 2016

Sajjad, Syed Muhammad, and Muhammad Yousaf "Security analysis of IEEE 802.15. 4 MAC in the context of Internet of Things (IoT)." In *2014 Conference on Information Assurance and Cyber Security (CIACS)*, pp. 9–14. IEEE, 2014.

Sepúlveda, Johanna, Shiyang Liu, and Jose M. Bermudo Mera "Post-quantum enabled cyber-physical systems." *IEEE Embedded Systems Letters* 11, no. 4 (2019): 106–110.

Shabana Kalsoom, Nigar Fida, Fazlullah Khan, Syed Roohullah Jan, and Mujeeb Ur Rehman "Security issues and attacks in wireless sensor networks." *International Journal of Advanced Research in Computer Science and Electronics Engineering (IJARCSEE)* 5, no. 7 (2016): 81–87.

Shahzad, Aamir, Malrey Lee, Young-Keun Lee, Suntae Kim, Naixue Xiong, Jae-Young Choi, and Younghwa Cho. "Real time MODBUS transmissions and cryptography security designs and enhancements of protocol sensitive information." *Symmetry* 7, no. 3 (2015): 1176–1210.

Sharma, Mridula, Fayez Gebali, Haytham Elmiligi, and Musfiq Rahman "Network security evaluation scheme for wsn in cyber-physical systems." In *2018 IEEE 9th Annual Information Technology, Electronics and Mobile Communication Conference (IEMCON)*, pp. 1145–1151. IEEE, 2018.

Sklavos, Nicolas, and Ioannis D. Zaharakis. "Cryptography and security in internet of things (IoTs): Models, schemes, and implementations." In *2016 8th IFIP International Conference on New Technologies, Mobility and Security (NTMS)*, pp. 1–2. IEEE, 2016.

Siddharth, Sridhar, Adam Hahn, and Manimaran Govindarasu "Cyber–physical system security for the electric power grid." *Proceedings of the IEEE* 100, no. 1 (2011): 210–224.

Stouffer, Keith, Joe Falco, and Karen Scarfone "Guide to Industrial Control Systems (ICS) security." *NIST Special Publication* 800 (2007): 82.

Summerville, Douglas H., Kenneth M. Zach, and Yu Chen "Ultra-lightweight deep packet anomaly detection for Internet of Things devices." In *2015 IEEE 34th International Performance Computing and Communications Conference (IPCCC)*, pp. 1–8. IEEE, 2015.

Thanigaivelan, Nanda Kumar, Ethiopia Nigussie, Rajeev Kumar Kanth, Seppo Virtanen, and Jouni Isoaho "Distributed internal anomaly detection system for Internet-of-Things." In *2016 13th IEEE annual consumer communications & networking conference (CCNC)*, pp. 319–320. IEEE, 2016.

Tsang Rose "Cyberthreats, vulnerabilities and attacks on scada networks." *University of California, Berkeley, Working Paper*, http://gspp.berkeley.edu/iths/TsangSCADA%20Attacks.pdf (as of Dec. 28, 2011) (2010).

Turk, Robert J. *Cyber incidents involving control systems.* No. INL/EXT-05-00671. Idaho National Laboratory (INL), 2005.

Vegh, Laura, and Liviu Miclea "Secure and efficient communication in cyber-physical systems through cryptography and complex event processing." In *2016 International Conference on Communications (COMM)*, pp. 273–276. IEEE, 2016.

Vacca, John R. *Computer and Information Security Handbook.* Newnes, 2012.

Wang, Y., Ruan, D., Gu, D., Gao, J., Liu, D., Xu, J., Chen, F., Dai, F. & Yang, J. Analysis of Smart Grid security standards. *2011 IEEE International Conference on Computer Science And Automation Engineering.* Vol. 4, pp. 697–701. 2011.

Wood, A. D. and Stankovic, J. A. "Denial of service in sensor networks," *Computer*, vol. 35, no. 10, pp. 54–62, 2002.

Wurm Jacob, Yier Jin, Yang Liu, Shiyan Hu, Kenneth Heffner, Fahim Rahman, and Mark Tehranipoor. "Introduction to cyber-physical system security: A cross-layer perspective." *IEEE Transactions on Multi-Scale Computing Systems* 3, no. 3 (2016): 215–227.

Xu, Kui, Ke Tian, Danfeng Yao, and Barbara G. Ryder "A sharper sense of self: Probabilistic reasoning of program behaviors for anomaly detection with context sensitivity." In *2016 46th Annual IEEE/IFIP International Conference on Dependable Systems and Networks (DSN)*, pp. 467–478. IEEE, 2016.

Yang, H., H. Luo, F. Ye, S. Lu, and L. Zhang "Security in mobile ad hoc networks: Challenges and solutions." *IEEE Wireless Communications* 11, no. 1, (2004): 38–47.

Ye, Haina, Xinzhou Cheng, Mingqiang Yuan, Lexi Xu, Jie Gao, and Chen Cheng "A survey of security and privacy in big data." In *2016 16th International Symposium on Communications and Information Technologies (ISCIT)*, pp. 268–272. IEEE, 2016.

Yemini, M., A. Nedic, A. Goldsmith, and S. Gil "Characterizing trust and resilience in distributed consensus for cyberphysical systems." *IEEE Transactions On Robotics* 38, (2021): 1–21.

Zarpelão, Bruno Bogaz, Rodrigo Sanches Miani, Cláudio Toshio Kawakani, and Sean Carlisto de Alvarenga "A survey of intrusion detection in Internet of Things." *Journal of Network and Computer Applications* 84 (2017): 25–37.

Zeadally, Sherali, Ray Hunt, Yuh-Shyan Chen, Angela Irwin, and Aamir Hassan. "Vehicular ad hoc networks (VANETS): status, results, and challenges." *Telecommunication Systems* 50, no. 4 (2012): 217–241.

Zhou, Tianqi, Jian Shen, Xiong Li, Chen Wang, and Haowen Tan "Logarithmic encryption scheme for cyber–physical systems employing Fibonacci Q-matrix." *Future Generation Computer Systems* 108 (2020): 1307–1313.

Znaidi Wassim, Marine Minier, and Jean-Philippe Babau. "An ontology for attacks in wireless sensor networks." PhD diss., INRIA, 2008.

8

Deep Learning Methods to Resolve Cyber Security Issues

Somenath Chakraborty

The University of Southern Mississippi, Hattiesburg, MS, USA

CONTENTS

8.1 Introduction

Cyber security consists of two terms. One is Cyber and the other one is Security. Cyber signifies any kind of computing or digital device. Devices used in communication channels have some analog devices in the network but the integrated architecture must have controlled devices which are nowadays in digital devices. Even ingenious electronics devices, home appliances, embedded electrical and electronics devices, IoT devices, and all kinds of smart devices use some kind of an integrated circuit to make the control architecture that comes under the cyber-physical work. The bigger version is big datacenters, private networks, patched digital hardware, supercomputers, mainframe, and all kinds of servers that are part of cyber-physical systems. Now, the second term of 'cyber security' is security. Security signifies a shield of protection from the outer world or securing the bridges of information from its network or within the devices.

In Figure 8.1 we see how many kinds of smart home appliances are used in the current scenario. The growth of the industry in this sector alone is very promising. Every smart device can be controlled by smartphones or can be controlled remotely. It has its own IP and other digital footprints. In this digital world, cyber security is not restricted only in the internet security aspects but 'Cyber Security' becomes an umbrella term and associated with any kinds of CPS. Figure 8.2. shows how artificial intelligence (AI) can provide key solutions to the broad variety of problem domains in the field of network venerability

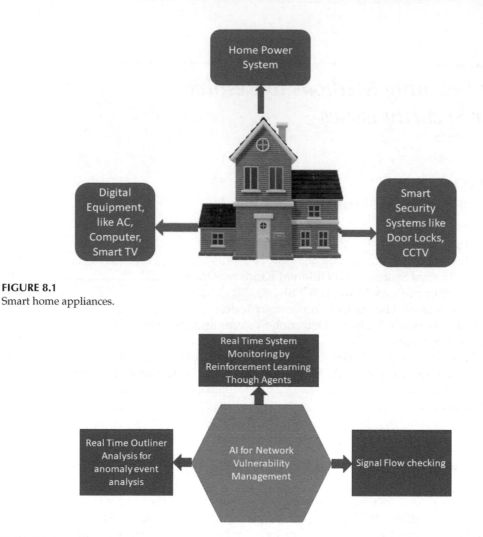

FIGURE 8.1
Smart home appliances.

FIGURE 8.2
AI for network vulnerability management.

management. It shows mainly three key aspects we need to focus on; those are as follows:

a. Real-time system monitoring by reinforcement learning through automated agents.
b. Real-time outliner analysis for anomaly event analysis.
c. Signal flow checking of all the connected computing nodes within a system.

Currently, the most active branches of Cyber Security fields are as follows:

a. Internet security
b. Web application and smartphone application-level security
c. Computer network and infrastructure level security
d. Intrusion detection and penetration testing

   e. Digital forensics and incident response-level security

   f. Endpoint protection and smartphone security

   g. Data governance, risk, and compliance

   h. Cloud computing security

   i. Power station and power grid security

   j. Centralized, decentralized, and distributed computing security

   k. Application layer security for home, car, office, and security appliances

   l. Integrated security for IoT devices

8.2 Cyber-Physical Systems and Different Kinds of Attacks

8.2.1 Cyber-physical Systems (CPS)

The computer network and distributed computing evolved a lot over the years. It consists of a hardware networking device, physical cables, different kinds of sensors, and many more. This hugged, gigantic physical structure and communications systems as an integrated control structure is regarded as CPS. In this domain of research, the CPS model security is well represented by M. Burmester et al. [1]. The stupendous, mammoth CPS with all its large-scale heterogeneous systems and devices is not fully perfect. There are lots of software, hardware, networking, and communication loophole and backdoor exists which makes CPS susceptible to different kinds of attack. Some infamous recent attacks are the Stuxnet worm creation, which takes control of the programmable logic controllers (PLCs) of nuclear centrifuges in Iran [2]. Later, these kinds of Stuxnet worms spread across the globe, and currently, more than 123 countries were affected by those kinds of the worm. Another one is the attack on a sewage treatment facility in Queensland, Australia, which manipulated the SCADA system to release raw sewage into local rivers and parks [3] (Figure 8.3).

If I talk about the recent attacks in the US alone, then there is a 300% increase in the payments made by the companies and growing, according to Harvard Business Review [4].

Many of these cyber-attacks use sophisticated hardware and software technology. Cyber-attacks in 2021 that have used ransomware as their attack vector include attacks made against the Colonial Pipeline, Steamship Authority of Massachusetts, the world's largest meatpacker JBS Foods, and the Washington DC Metropolitan Police Department. These attacks against US companies and organizations result in a shutdown of critical infrastructure, create shortages in necessary food supply and oil, increased cost of goods and services, different kinds of financial loss as many companies are forced to make partial and complete shutdown of their operation due to risk factors, even loss of money due to having to pay the ransom to the hackers.

The different kinds of cyber-attacks are as follows:

   a. Malware-related attacks

   b. Phishing attacks

   c. Password theft-related attacks

   d. Distributed denial-of-service (DDoS) attacks

   e. Man in the middle attacks

   f. Drive-by download attacks

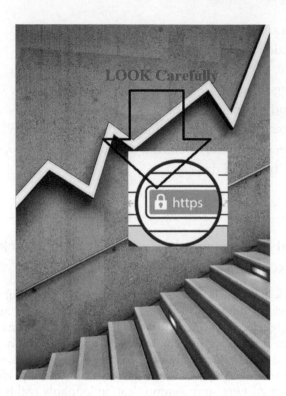

FIGURE 8.3
Just by checking into secure https before browsing saves millions of online attacks.

 g. Fake advertising–related attacks

 h. Rogue software attacks

Now, the chapter would introspect some of the attacks in detail. Then, it would present both the traditional approaches and machine learning approaches to prevent and design protected systems with more robustness.

8.3 Malware-related Attacks and Prevention Models

Malware attacks are the most common types of attacks in cyber security. Malware is a special kind of bug software like virus software, spyware software, adware software, worms software, Trojan horse software, botnets software, etc., which potentially harm computing systems by exploiting their vulnerability. Sometimes, it creates different kinds of computing hazards to the system and sometimes steals valuable information of the targeted system. These types of software usually self-replicate and spread quickly throughout the computing system and network. The difference between worms and virus is that worms can self-propagate and damage the system without the user-end initiative, whereas virus uses some kinds of executable file which looks like authenticate to the user and the user initiates the program in his or her devices and then it starts to propagate and self-replicate through computing devices and networks. Currently, some advanced kinds of malware

are able to camouflage very authentic software rather than an executable program which is hard to detect, and traditional malware detection techniques failed to detect those. According to the Cisco/Cybersecurity Ventures 2019 Cybersecurity Almanac [5], approximately 1 million malware files are created every day, and cybercrime will damage the world economy by approximately $6 trillion annually by 2021. Table 8.1. presents different kinds of malware detection approaches where traditional approaches are used.

Traditional malware detection technique mostly depends on the analysis which is not able to shield the systems. Mostly, they used supervised techniques and heuristic

TABLE 8.1

Traditional Malware Detection Approaches

Research Work	Method Overview	Objectives	Year of Inclusion
T.F. Lunt [6]	Automated audit trail analysis techniques.	Intrusion detection	1988
S. Axelsson [7]	Taxonomy and survey	Intrusion detection both formal and informal ways	2000
Schultz et al. [8]	Generate strings using DLL and API system calls	Malware detection	2001
Sing et al. [9]	LTL mechanism	Explore different variety of malware	2003
Kinder et al. [10]	Tree validation logic	Explore different variety of malware	2005
Zhang et al. [11]	Ensemble learning approach of multiple classifiers	Signature-based technique for detecting viruses	2007
Karnik et al. [12]	Cosine similarity analysis	Signature-based virus detection	2007
Moser et al. [13]	Execution path detection analysis	Malware detection	2007
Martignoni et al. [14]	Behavior-based analysis	Malware analysis for cloud systems	2009
Griffin et al. [15]	Automatic generation of string signatures	Explore different variety of malware	2009
Anderson et al. [16]	Graph-based analysis	Explore different variety of malware	2011
Cha et al. [17]	Distributed approach	Signature-based malware detection	2011
Isohara et al. [18]	Kernel-based behavior analysis	Android malware detection	2011
Shabtai et al. [19]	Host-based feature monitoring system	Android malware detection	2012
Song et al. [20]	Model analysis	Malware detection	2012
Baldangombo et al. [21]	Data mining method	Signature-based malware detection	2013
Park et al. [22]	Graph clustering approach	Behavior-based malware detection	2013
Islam et al. [23]	Integrated static and dynamic feature analysis	Different types of malware detection	2013
Naval et al. [24]	Asymptotic equipartition property (AEP) analysis	Evasion–proof Malware detection	2015
Hao et al. [25]	Suspicious bucket filtering mechanism (RScam)	Cloud-based malware detection	2015
Huang et al. [26]	Multitask neural network	Dynamic malware detection	2016
Norouzi et al. [27]	Datamining classification approach	Behavior-based malware detection	2016
Das et al. [28]	Semantics-based online detection	Behavior-based malware detection	2016
Aashima et al. [29]	Hybrid pattern based text mining approach	Malware detection (DBScam)	2016
Cimitile et al. [30]	Phylogenetic trees checking-based method	Android malware evolution	2017
Jyothsna et al. [31]	Meta-heuristic assessment model to assess degree of intrusion scope	Intrusion detection technique	2018

techniques which mostly depend on past databases and feature format. The main drawback of this system is that it is not fully real-time based and the previously used database is not able to provide dynamic behavior of the system in a precious definite way. Because of the urgency of the need, new techniques are investigated. After 2016, deep learning, reinforcement learning, blockchain techniques become so popular that almost every sector get benefited from this. Cyber security is also no exception in this regard. Different kinds of cyber security approaches are integrated, partially used, and sometimes totally replaced the old approaches to explore the vulnerability of the system and build a shielded system to protect the system by the deep learning mechanism.

Table 8.2. investigates and presents all new kinds of machine learning technique which are introduced in recent era.

TABLE 8.2

Deep Learning Methods for Malware Detection

Research Work	Method Overview	Objectives	Year of Inclusion
Zhu et al. [32]	Presents a DBN model, DeepFlow technique for analyzing hybrid features	Deep learning approach to malware detection	2017
Xiao et al. [33]	Q-learning, post-decision state learning-based scheme	Cloud-based malware detection for mobile devices	2017
Narayanan et al. [34]	Four-stage analysis framework, static analysis, feature extraction and representation, online learning and explanation	Online learning-based framework for malware detection, they named it as CASANDRA (Context-aware, Adaptive and Scalable ANDRoid mAlware detector).	2017
Lee et al. [35]	Stacked convolutional and recurrent neural networks (CNN, RCN), they named it as SeqDroid	Obfuscated Android malware detection	2019
Wang et al. [36]	Deep autoencoder and convolutional neural network	Android malware detection	2019
Qin et al. [37]	Multiclass features and deep belief network, they name it as MSNdroid	Android malware detection	2019
Sun et al. [38]	Deep learning–based approach using code images	Android malware classification	2019
Zou et al. [39]	Deep learning–based approach using byte-code sequences, they named it as ByteDroid	Android malware detection	2019
Chen et al. [40]	Word2Vec and deep belief network, they named it as DroidvecDeep	Android malware detection	2019
Zhu et al. [41]	CNN-based approach using backtracking method to infer suspicious features of the apps like permission, API, URL, etc. to explain the reason of classification	Android malware detection	2019
Ren et al. [42]	CNN and long short-term memory (LSTM)	Malware detection for android IoT devices	2020
Alzaylaee et al. [43]	Deep learning system (DL-Droid) to detect malicious Android applications, experiment performed on 30,000 applications	Dynamic analysis using stateful input generation	2020
Li et al. [44]	Opcode sequence analysis using CNN	Android malware detection	2020
Su et al. [45]	Deep belief network (DroidDeep) using 11 different kinds of static behavioral characteristics to analyze and classify large amount of android applications	Android malware detection	2020

8.4 Difference between Malware Detection Systems and Intrusion Detection Systems

Although these two terms come across every time, we introspect any computer system but there are certain differences that exist in these two terms.

The term malware is more specific than an intrusion. It covers a range of defined categories such as viruses, worms, trojan horse, and ransomware. Intrusion, on the other hand, is much broader in the sense and used in a holistic way. It also includes an active actor such as an individual behind numerous programs systematically trying to gain access to a computing node or network.

When talking about malware, whether anomaly- or signature-based detection techniques, there is reliance that some parts of the malware are static for detection. That is, there is a component that is predictable and can be used for detection. The easiest way to understand this is with signature detection. It relies on identifying a part of program, add that particular feature to the malware detector, and deploy the solution.

Intrusion detection systems can also be malware detectors because they use signature and behavior-based techniques. The difference is in the scope of what is being targeted. With malware, the behavior is more deterministic; with intrusion it is broader and requires further interpretation and cannot trust on one set of approaches rather than a branch of integrated approaches or detection mechanism. In this sense, intrusion detection is more amenable to behavior-based profile. If some predefined broad rules are triggered, then inspection is merited. That is, an alarm trigger may be malware but also an active intrusion by an actor and it is up to a person or program to try to sort it out.

In a nutshell, a malware program is more specific than an intrusion detection. An easy way to separate them is on the scope of what they are targeting. Be mindful that it may not be easy to draw the distinction with a particular technique or program. It all depends on the particular implementation and what the software is targeting.

8.5 Phishing Attacks and Prevention Systems Using Machine Learning

Phishing attacks are mostly happen with email spoofing. Sometimes, it looks authentic to the victims or any user.

Figure 8.4. gives the general overview of the three-step process involved in phishing attacks. Initially, the user receives many emails which have many suspicious fake website links that somehow disguise in such a way that users feel tempted with great offers or potential benefits and they click on that link. Then, a fake website pops up and it looks very authentic, it contains form fill-up fields, and the user provides their valuable information. Then, the online predators or phishing attackers steal the information and sell them or use that information for their own benefits.

There is much research exploring machine learning's potential to tackle the problem in a more efficient manner.

Supervised, semi-supervised, unsupervised, and reinforcement algorithms like radial basis function network (RBFN), naive Bayes classifier (NB), decision tree, back-propagation neural network (BPNN), k-Nearest neighbor (kNN), random forest (RF), support vector machine (SVM), statistical mixed and meta models, and deep learning are used in different literature. Table 8.3. presents the recent machine learning approaches in this regard.

FIGURE 8.4
General overview of the steps in Phishing attacks.

TABLE 8.3

Machine Learning Methods for Preventing Phishing Attacks and Detecting Phishing Websites

Research Work	Method Overview	Objectives	Year of Inclusion
Abdelhamid et al. [46]	They investigate a large number of websites with 30 features with the help of PHP script that was embedded in the web browser and applied on websites sources.	Analysis to classify websites as a phishing website or not.	2017
Harikrishnan et al. [47]	Distributional representation, namely TF-IDF for numeric representation approach to classify phishing emails	Phishing email detection	2018
Vinayakumar et al. [48]	Classify domain generation algorithms (DGAs) to detect phishing websites using deep learning methods	Detect and classify the pseudo random domain names or phishing websites without relying on the feature engineering or any other linguistic, contextual, or semantics and statistical information	2018
Fang et al. [49]	A modified recurrent convolutional neural networks (RCNN) model with multilevel vectors and attention mechanism, they named their model as THEMIS	Phishing email detection	2019
Alsariera et al. [50]	AI meta-learners and extra-trees algorithm	Detection of phishing websites	2020
Sameen et al. [51]	Ensemble machine learning-based detection system, they named it as PhishHaven to identify AI-generated as well as human-crafted phishing URLs	Phishing URLs detection system	2020
Indrasiri et al. [52]	Robust ensemble machine learning model using expandable random gradient stacked voting classifier (ERG-SVC)	Detecting phishing websites	2021

Also, this chapter introspects the research work done by Chen et al. [53] to demonstrate how to generate phishing apps (phapps) automatically. They observe that initiating a phishing attack requires two conditions, page confusion and logic deception, during attack synthesis. They optimize these to create practical phishing attacks. Their experimental results reveal that existing phishing defenses are less effective against such emergent attacks. Also, newly proposed attacks still remain mostly undetected and are worth further exploration.

8.6 Critical Comparison of the AI Methodologies with the Traditional Systems

There are many aspects that deep learning techniques address over the past few years in the context of addressing cyber security issues. One of the fundamental changes in the real-time implementation of the model boosts the detection cycle of the network. Especially the vulnerability is now detected by the automated agent monitoring system all the time. Another important contribution to the introduction of an automated system is speedy detection and recovery management through distributed redundancy of critical resources. Online cloud management fills up the gaps of critical resource-related problems and venerability issues. Overall, the performance enhances a lot as many bugs that were previously undetected consume a lot of resources and that sometimes became the cause of resource failure which are nowadays easily avoided while integrating our system with the state-of-the-art deep learning security methods.

8.7 Conclusion

This chapter explores different cyber security aspects and their corresponding research emphasizing machine learning's perspective in detail. The comprehensive study mainly focuses on malware detection and malware prevention system design, intrusion detection system design, and phishing attacks–related research in the context of recent developments, machine learning, and AI technique, so that it becomes more robust, reliable, real-time, and cost-effective. In the future, other types of cybersecurity aspects would be investigated and present new kinds of research which could show the solution aspects.

References

1. M. Burmester, E. Magkos, and V. Chrissikopoulos. Modeling security in cyber-physical systems. IJCIP, 5(3–4):118–126, 2012.
2. N. Falliere, L. Murchu, and E. Chien. W32.Stuxnet Dossier, 2011.
3. J. Slay and M. Miller. Lessons Learned from the Maroochy Water Breach. In E. Goetz, S. Shenoi (Eds.), *Critical Infrastructure Protection, IFIP 253*, pages 73–82. Springer, 2007.
4. Harvard Business Review newsletters published in the series of Cyber Security, 2020, 2021.

5. S. Morgan. 2019/2020 Cybersecurity Almanac: 100 Facts, Figures, Predictions And Statistics. Accessed: Dec. 5, 2021. [Online]. Available: https://cybersecurityventures.com/cybersecurity-almanac-2019.

6. T. F. Lunt, Automated audit trail analysis and intrusion detection: a survey, in *Proceedings of the 11th National Computer Security Conference*, 1988, vol. 353: Baltimore, MD.

7. S. Axelsson, Intrusion detection systems: a survey and taxonomy, technical report 2000.

8. M. G. Schultz, E. Eskin, F. Zadok, and S. J. Stolfo, Data mining methods for detection of new malicious executables, in *Proc. IEEE Symp. Secur.* Privacy, May 2001.

9. P. Singh and A. Lakhotia, Static verification of worm and virus behavior in binary executables using model checking, in *Proc. IEEE Syst., Man Soc. Inf. Assurance Workshop*, Mar. 2003.

10. J. Kinder, S. Katzenbeisser, C. Schallhart, and H. Veith, Detecting malicious code by model checking, in *Proc. Int. Conf. Detection Intrusions Malware, Vulnerability Assessment*. Berlin, Germany: Springer, 2005.

11. B. Zhang, J. Yin, J. Hao, D. Zhang, and S. Wang, Malicious codes detection based on ensemble learning, in *Autonomic and Trusted Computing (Lecture Notes in Computer Science)*, vol. 4610. Berlin, Germany: Springer, 2007, pp. 468–477.

12. A. Karnik, S. Goswami, and R. Guha, Detecting obfuscated viruses using cosine similarity analysis, in *Proc. 1st Asia Int. Conf. Modeling Simulation (AMS)*, Mar. 2007.

13. A. Moser, C. Kruegel, and E. Kirda, Exploring multiple execution paths for malware analysis, in *Proc. IEEE Symp. Secur.* Privacy (SP), May 2007.

14. L. Martignoni, R. Paleari, and D. Bruschi, A framework for behavior based malware analysis in the cloud, in *Proc. Int. Conf. Inf. Syst. Security*. Berlin, Germany: Springer, 2009.

15. K. Griffin, S. Schneider, X. Hu, and T.C. Chiueh, Automatic generation of string signatures for malware detection, in *Proc. Int. Workshop Recent Adv. Intrusion Detection*. Berlin, Germany: Springer, 2009.

16. B. Anderson, D. Quist, J. Neil, C. Storlie, and T. Lane, Graph-based malware detection using dynamic analysis, J. Comput. Virol., vol. 7, no. 4, pp. 247–258, Nov. 2011.

17. S. K. Cha, I. Moraru, J. Jang, J. Truelove, D. Brumley, and D. G. Andersen, SplitScreen: Enabling efficient, distributed malware detection, J. Commun. Netw., vol. 13, no. 2, pp. 187–200, Apr. 2011.

18. T. Isohara, K. Takemori, and A. Kubota, Kernel-based behavior analysis for Android malware detection, in *Proc. 7th Int. Conf. Comput. Intell. Secur.*, Dec. 2011.

19. A. Shabtai, U. Kanonov, Y. Elovici, C. Glezer, and Y. Weiss, Andromaly: A behavioral malware detection framework for Android devices, J. Intell. Inf. Syst., vol. 38, no. 1, pp. 161–190, Feb. 2012.

20. F. Song and T. Touili, Efficient malware detection using model checking, in *Proc. Int. Symp. Formal Techn.* Berlin, Germany: Springer, 2012.

21. U. Baldangombo, N. Jambaljav, and S.-J. Horng, A static malware detection system using data mining methods, 2013, arXiv:1308.2831. [Online]. Available: https://arxiv.org/abs/1308.2831.

22. Y. Park, D. S. Reeves, and M. Stamp, Deriving common malware behavior through graph clustering, Comput. Secur., vol. 39, pp. 419–430, Nov. 2013.

23. R. Islam, R. Tian, L. M. Batten, and S. Versteeg, Classification of malware based on integrated static and dynamic features, J. Netw. Comput. Appl., vol. 36, no. 2, pp. 646–656, Mar. 2013.

24. S. Naval, V. Laxmi, M. Rajarajan, M. S. Gaur, and M. Conti, Employing program semantics for malware detection, IEEE Trans. Inf. Forensics Security, vol. 10, no. 12, pp. 2591–2604, Dec. 2015.

25. H. Sun, X. Wang, J. Su, and P. Chen, RScam: Cloud-based anti-malware via reversible sketch, in *Proc. Int. Conf. Secur. Privacy Commun. Syst.* Cham, Switzerland: Springer, 2015.

26. W. Huang and J. W. Stokes, MtNet: A multi-task neural network for dynamic malware classification, in *Proc. Int. Conf. Detection Intrusions Malware, Vulnerability Assessment*. Cham, Switzerland: Springer, 2016.

27. M. Norouzi, A. Souri, and S. M. Zamini, A data mining classification approach for behavioral malware detection, J. Comput. Netw. Commun., vol. 2016, p. 1, Mar. 2016.

28. S. Das, Y. Liu, W. Zhang, and M. Chandramohan, Semantics based online malware detection: Towards efficient real-time protection against malware, IEEE Trans. Inf. Forensics Security, vol. 11, no. 2, pp. 289–302, Feb. 2016.

29. A. Malhotra and K. Bajaj, A hybrid pattern based text mining approach for malware detection using DBScan, CSI Trans., vol. 4, nos. 2–4, pp. 141–149, Dec. 2016.

30. A. Cimitile, F. Martinelli, F. Mercaldo, V. Nardone, A. Santone, and G. Vaglini, Model checking for mobile android malware evolution, in *Proc. IEEE/ACM 5th Int. FME Workshop Formal Methods Softw. Eng. (FormaliSE)*, May 2017.

31. Jyothsna, V., Rama Prasad, V.V. Assessing degree of intrusion scope (DIS): a statistical strategy for anomaly based intrusion detection. CSIT 6, 99–127 (2018). https://doi.org/10.1007/s40012-018-0188-x.

32. D. Zhu, H. Jin, Y. Yang, D. Wu, and W. Chen, DeepFlow: Deep learning based malware detection by mining Android application for abnormal usage of sensitive data, in Proc. IEEE Symp. Comput. Commun. (ISCC), Jul. 2017.

33. L. Xiao, Y. Li, X. Huang, and X. Du, Cloud-based malware detection game for mobile devices with offloading, IEEE Trans. Mobile Comput., vol. 16, no. 10, pp. 2742–2750, Oct. 2017.

34. Narayanan, Annamalai, Mahinthan Chandramohan, Lihui Chen and Yang Liu. Context-Aware, Adaptive, and Scalable Android Malware Detection Through Online Learning. IEEE Transactions on Emerging Topics in Computational Intelligence 1 (2017): 157–175.

35. W. Y. Lee, J. Saxe, and R. Harang, SeqDroid: Obfuscated Android malware detection using stacked convolutional and recurrent neural networks, in Mamoun Alazab, Ming Jian Tang (Eds.), Deep Learning Applications for Cyber Security. Springer, 2019, pp. 197–210.

36. W. Wang, M. Zhao, and J. Wang, Effective Android malware detection with a hybrid model based on deep autoencoder and convolutional neural network, J. Ambient Intell. Humanized Comput., vol. 10, no. 8, pp. 3035–3043, Aug. 2019.

37. X. Qin, F. Zeng, and Y. Zhang, MSNdroid: The Android malware detector based on multi-class features and deep belief network, in *Proc. ACM Turing Celebration Conf.-China*, 2019, pp. 1–5.

38. Y. Sun, Y. Chen, Y. Pan, and L. Wu, Android malware family classification based on deep learning of code images, IAENG Int. J. Comput. Sci., vol. 46, no. 4, 2019.

39. K. Zou, X. Luo, P. Liu, W. Wang, and H. Wang, ByteDroid: Android malware detection using deep learning on bytecode sequences, in *Proc. Chin. Conf. Trusted Comput. Inf. Secur.* Springer, 2019, pp. 159–176.

40. T. Chen, Q. Mao, M. Lv, H. Cheng, and Y. Li, DroidvecDeep: Android malware detection based on Word2Vec and deep belief network, TIIS, vol. 13, no. 4, pp. 2180–2197, 2019.

41. D. Zhu, T. Xi, P. Jing, D. Wu, Q. Xia, and Y. Zhang, A transparent and multimodal malware detection method for Android apps, in *Proc. 22nd Int. ACM Conf. Modeling, Anal. Simulation Wireless Mobile Syst.*, 2019, pp. 51–60.

42. D. Li, L. Zhao, Q. Cheng, N. Lu, and W. Shi, Opcode sequence analysis of Android malware by a convolutional neural network, Concurrency Comput. Pract. Exper., vol. 32, no. 18, p. e5308, 2020.

43. Z. Ren, H. Wu, Q. Ning, I. Hussain, and B. Chen, End-to-end malware detection for Android IoT devices using deep learning, Ad Hoc Netw., vol. 101, Apr. 2020, Art. no. 102098.

44. M. K. Alzaylaee, S. Y. Yerima, and S. Sezer, Dl-Droid: Deep learning based Android malware detection using real devices, Comput. Secur., vol. 89, Feb. 2020, Art. no. 101663.

45. X. Su, W. Shi, X. Qu, Y. Zheng, and X. Liu, DroidDeep: Using deep belief network to characterize and detect Android malware, Soft Comput., Vol. 24, pp. 1–14, 2020.

46. Abdelhamid, Neda, Fadi A. Thabtah and Hussein Abdel-Jaber, Phishing detection: A recent intelligent machine learning comparison based on models content and features, *2017 IEEE International Conference on Intelligence and Security Informatics (ISI)* (2017): 72–77.

47. Harikrishnan, N. B., R. Vinayakumar, K. P. Soman and Das A. Verma R.M., A machine learning approach towards phishing email detection, *CEN-Security@IWSPA 2018.* (2018).

48. R Vinayakumar, KP Soman, Prabaharan Poornachandran, and S Sachin Kumar. Evaluating deep learning approaches to characterize and classify the DGAs at scale. Journal of Intelligent & Fuzzy Systems, 34(3):1265–1276, 2018.

49. Y. Fang, C. Zhang, C. Huang, L. Liu and Y. Yang, Phishing Email Detection Using Improved RCNN Model With Multilevel Vectors and Attention Mechanism, in IEEE Access, vol. 7, pp. 56329–56340, 2019, doi: 10.1109/ACCESS.2019.2913705.

50. Y. A. Alsariera, V. E. Adeyemo, A. O. Balogun and A. K. Alazzawi, AI Meta-Learners and Extra-Trees Algorithm for the Detection of Phishing Websites, in IEEE Access, vol. 8, pp. 142532–142542, 2020, doi: 10.1109/ACCESS.2020.3013699.

51. M. Sameen, K. Han and S. O. Hwang, PhishHaven—An Efficient Real-Time AI Phishing URLs Detection System, in IEEE Access, vol. 8, pp. 83425–83443, 2020, doi: 10.1109/ACCESS.2020.2991403.

52. P. L. Indrasiri, M. N. Halgamuge and A. Mohammad, Robust Ensemble Machine Learning Model for Filtering Phishing URLs: Expandable Random Gradient Stacked Voting Classifier (ERG-SVC), in IEEE Access, vol. 9, pp. 150142–150161, 2021, doi: 10.1109/ACCESS.2021.3124628.

53. S. Chen, L. Fan, C. Chen, M. Xue, Y. Liu and L. Xu, GUI-Squatting Attack: Automated Generation of Android Phishing Apps, in IEEE Transactions on Dependable and Secure Computing, vol. 18, no. 6, pp. 2551–2568, 1 Nov.–Dec. 2021, doi: 10.1109/TDSC.2019.2956035.

9

Application of Temporal Logic for Construction of Threat Models for Intelligent Cyber-Physical Systems

Manas Kumar Yogi and A. S. N. Chakravarthy

JNTUK, Kakinada, India

CONTENTS

9.1 Introduction

As technological innovations are increasing their role in human life, the advent of cyber-physical systems (CPS) is also enhancing their participation in every sphere of life. In all smart systems in modern world, the concepts of intelligent entities are related to how diverse the applications are made with principles of artificial intelligence and machine learning (Gritzalis, et al., 2019). Currently, we can say we are operating in the CPS 1.0 ecosystem, and by 2030, we will move toward CPS 2.0 ecosystem. Smart cities, smart vehicles, and smart industrial manufacturing units are all part of CPS 1.0 ecosystem (Keliris, et al., 2016). So, the question is what will be included as functional units as part of CPS 2.0 ecosystems. We can propagate the idea that CPS 2.0 will constitute more degree of interaction between heterogeneous components using mobile computing models and cloud computing models (Konstantinou, 2021). Spatial-temporal properties will be explored to a greater extent, and the environment, while execution will become more open. The paradigm of cyber-physical system of systems will become more in use rather than calling it as just cyber-physical systems (Muyeen and Rahman, 2017).

Intelligent CPS will incorporate two major features. One is self-adjustment and the other one is auto-optimization (Ospina, et al, 2021). Self-adjustment is a dynamic behavior and it will need substantial amount of data to perform adjustments according to a desired context. To achieve this, the designers of the intelligent CPS should follow principles of context-aware computing to render the self-adjustment property. Researchers in this area are faced with challenges in this because of lack of standard rules to generate the context-aware principles (Chen, et al., 2014). The other factor of auto-optimization can be achieved by various mathematical models of formalism (Queiroz, et al., 2011). These models of formalism are advocated as robust models of construction in theory, but in practical sense, they are more challenging to implement. Many designers have also applied bio-inspired models for auto-optimization but their computational complexity is still a hindrance to easily accept their strength (Dorsch, et al., 2014). The degree of insight presently installed in our vehicles, homes, specialized gadgets, shopper hardware, and different gadgets expands each day. In the extremely not so distant future, not exclusively will people connect with a quickly developing exhibit of smart items, yet large numbers of these items will interface independently with one another and different frameworks.

In addition, industrial facility creation lines, measure plants for energy and utilities, and smart urban communities will rely upon cyber-actual frameworks (CPS) to self-screen; upgrade; and surprisingly independently run foundation, transportation, and structures (Georg, et al., 2013). Later on, cyber-actual frameworks will depend less on human control and more on the knowledge implanted in the man-made reasoning (AI)-empowered center processors.

Because of the dramatic development in utilization of CPS and the ideal opportunity to showcase, the security dangers are not being considered as the principal configuration challenges in the CPS configuration cycle. However, a few safety efforts are being presented, the greater part of them are engaged toward the cyber-attacks (Stanovich, et al., 2013). Be that as it may, the purposeful (vindictive goal) or accidental (normal or ecological fiasco) physical attacks represent a few basic plan concerns. In outcomes, a key question emerges that how to insert the actual security measures during the CPS configuration cycle? Normally, the CPS are being tried furthermore and examined for the security weaknesses during the configuration stage or before the organization stage (post-manufacture testing stage). Be that as it may, a few unexpected (normal, natural, or coincidental fiascos) and deliberate dangers can happen during the runtime (Boschert and Rosen, 2016). To apply certain safety efforts during the runtime, it is urgent to present the runtime discovery and choice capacity in the CPS. Nonetheless, because of asset limitations and energy financial plan, it restricts the extent of runtime safety efforts which brings up an essential examination issue that how to plan a certain runtime security measure with greatest inclusion of safety weaknesses while considering the asset requirements and spending plan?

Qualitative assessments for cybersecurity hazards require generous framework information on the CPS design and experience from the associations and gatherings directing the investigation (Zhou, et al., 2019). Then again, quantitative studies ascertain accurate danger scores, which help the prioritization and moderation systems. Other works utilizing recreation helped examinations to assess the comparing effect of cyber-attacks. In addition, analysts have considered powerfully adjusting hazard assessment models figuring the framework and assault sway development for the danger score computations. Recent works have proposed mixes of various danger techniques outfitting the upsides of more than one methodology and giving more reasonable assessments (Huang, et al., 2016).

While makers across all modern areas increasingly trying to satisfy the need for this developing "smart item" market, they face significant difficulties creating and fabricating these new and progressively more perplexing items and frameworks. These cyber-actual frameworks require tight coordination and joining between the computational (virtual) and the physical (ceaseless) universes (Ospina, et al., 2019).

In general, the attacker model is composed of the following major aspects: resources under control, knowledge of how to use the resource under control, and specificity, access. Attacks termed as white-box represent attacks in which a malicious intent or an adversary is equipped with maximum useful information regarding the system architecture, its internal working parameters, and different internal states of the system (Keliris, et al., 2018). A gray-box attack denotes attacks where partial knowledge of the system is intrinsic to the attacker. In black-box attacks, the adversary does not have any knowledge about the system but tries to attack the system by deploying confidence scores. If the attacker gains access to few resources under his control, he can easily compromise the system.

9.2 Motivation and Contributions

We will now discuss the process of how the attacker can probe the weakness of cyber-physical ecosystem. The major weakness in a CPS can be grouped into the following categories:

1. Network-related weakness:
   These weaknesses result due to compromising open-wired/wireless communication and connections, gaps in communication-stack (network/transport/application layer), attacks related to man-in-the-middle, spoofing, eavesdropping, back-doors, replay, denial-of-service (DoS)/ distributed denial-of-service (DDoS), packet manipulation attacks, and sniffing attacks (Xiang, et al., 2017).

2. Management-related weakness:
   It arises when the organization involved in CPS operations does not have a robust process or policies related to security guidelines.

3. Platform-related weakness:
   It is propelled by weakness in software configuration or hardware configuration or even database-related vulnerabilities.

Security concerns going from application climate and correspondence innovation ought to be tended to at the beginning phases of the plan. In addition, the intrinsic qualities and benefits of utilizing accessible organizations, like wireless sensor networks (WSN), next-generation networks, and the Internet, CPS are progressively confronting new security challenges, like getting conventions and setting up trust between CPS subsystems (Tian et al., 2020). A significant number of the figuring subsystems in CPS depend on components off the self (COTS) parts. The COTS parts give a critical degree of control, lower organization, and lower functional expenses in contrast with the customary merchant explicit restrictive and shut source frameworks. Be that as it may, this opens CPS to more weaknesses and dangers. For instance, modern control frameworks have been viewed as secure when not associated with the rest of the world, without considering insider assaults (Tu, et al., 2020). Subsequently, this demonstrates that the broad availability among digital and actual parts raises the significant issue of safety.

The failures in a CPS can be broadly divided into below two:

1. **Independent failures**: The independent failures are mostly hardware failures and they have to be replaced in case of major failure. In minor cases, they can be repaired or reconfigured.

2. **Correlated failures**: But unlike the above, correlated failures have multiple sources of failures as discussed below.

The risk of related disappointments turns out to be particularly significant in CPSs because of the tight coupling of normally ceaseless actual elements and discrete elements of installed registering measures (Pan, et al., 2017).

Correlated failures start from at least one of the accompanying occasions:

1. **Simultaneous assaults**: Targeted digital assaults (e.g., failures because of Stuxnet); non-focused on digital assaults (e.g., disappointments because of Slammer worm, conveyed DoS assaults, blockage in shared organizations); facilitated actual assaults (e.g., failures brought about by militants)

2. **Simultaneous deficiencies**: Common-mode disappointments (e.g., disappointment of various ICT parts in an indistinguishable way, programming blunders); arbitrary disappointments (e.g., regular occasions such as seismic tremors and typhoons, and administrator mistakes, for example, a mistaken firmware update)

3. **Cascading disappointments**: Failure of a negligible part of hubs (parts) in one CPS sub-network can prompt reformist acceleration of disappointments in other sub-networks (e.g., power outages in organizations influencing correspondence organizations).

Because of the restrictively significant expenses of data securing, it is frequently too exorbitant to even consider deciding the accompanying:

- Which equipment glitches and programming bugs have caused a framework disappointment?
- Whether the framework disappointment was brought about by an unwavering quality disappointment or security disappointment or both.

As a rule, these data fluctuate essentially across various elements (players), like CPS administrators, SCADA, and ICT merchants, network specialist organizations, clients, and neighborhood/bureaucratic administrative offices (or government) (Barua and Al Faruque, 2020). Data insufficiencies emerge from the clashing interests of individual players whose decisions influence the CPS chances. One might say that associated disappointments cause externalities that result in skewed player motivating forces (i.e., the independently ideal CPS security safeguards separate from the socially ideal ones). In addition, in conditions with deficient and furthermore awry (and private) data, the cultural expenses of an associated CPS disappointment normally surpass the misfortunes of the singular players whose items and administrations influence CPS activities, and on whose activities the CPS chances depend (Venkataramanan, et al., 2019). In particular, interdependencies among security and dependability disappointments in CPS are probably going to cause negative externalities. In such conditions, the singular players tend to underinvest in security comparative with a socially ideal benchmark (Zhang, et al., 2019). This requires plan of institutional means to realign the singular players' motivators to make sufficient interests in security. Instances of institutional means incorporate guidelines that expect players to guarantee that they have specific security capacities, and legitimate principles which command that players share data about security episodes with government offices or potentially the general population through set up channels.

9.3 Related Work

Threat modeling is a technique used by system or network designers to bring out the weak points in a system before the actual attack happens. After threat modeling is performed, mitigation strategies are developed to counter the threats. It is obvious that performing danger displaying for CPS is fundamental because their trade off can have tragic outcomes to the matrix activity and the financial and social prosperity. Notwithstanding, CPS comprise of different layers and resources; subsequently, it tends to be trying, because of broad time, demonstrating endeavors, assets, and cost, to thoroughly analyze every one of the potential situations that could emerge as framework weaknesses (Kuruvila, et al., 2020). To defeat such issues, without compromising the framework's dependability, numerous danger-demonstrating novel mechanisms have been put forth expecting to focus on weaknesses and help the execution of strong security components. These strategies give an all-encompassing perspective on the framework by featuring the huge resources, regularly alluded to as royal gems, and surveying dangers dependent on their possible effect and simplicity of sending on the framework.

Utilizing any of these threat modelling approaches draws near, risk is assessed by studying the popular threat contexts in connection with the pertinent weaknesses. The subsequent measure is the consequence of any intrinsic risk, defined as the moderation of threats given by executed controls, and comprises a proportion of the remaining risk. This interaction might repeat as extra controls are distinguished and carried out and as advancing threat abilities are recognized and detailed. Estimating risk levels and recognizing functional cycles that help continuous moderation of cyber threats should bring about an announcing capacity for huge risk-based measurements. Fault tree analysis (FTA) is the earliest innovation in safety risk assessment, and it is a graphical procedure broadly utilized for peril and risk assessment in CPS. The principle objective of FTA is to introduce the conceivable typical and faulty occasions that can cause the high-level undesired occasion. The fault tree comprises of the accompanying parts: hubs (undesired occasions in the framework), entryways (relations between hubs; it can be AND OR doors), and edges (way of the undesired occasions through the framework). The failure modes and effects analysis (FMEA) is an organized and group-based technique for framework safety analysis to perceive, assess, and score expected failures and their effects. Failure mode alludes to the manner by which something may come up short, and impact analysis is utilized to score the seriousness of different failure modes. The term risk priority number (RPN) is a piece of FMEA quantitative analysis; it is the result of the seriousness, likelihood of an event, and location likelihood. Model-based engineering (MBE) is a technique for creating conduct models of continuous frameworks and investigating the models for prerequisite confirmation to guarantee safety of CPS. To start with, the strategy considers the framework safety to decide a bunch of anticipated properties, then, at that point, it removes properties of the physical climate, figuring units and the cyber-physical connections, and lastly, investigations into the theoretical model to assess the normal properties and confirm safety necessities.

Improvement of measurements is outside the extent of this report, yet risk measurements are basic to furnishing chief supervisors with oversight abilities to set up a cyber-program benchmark to oversee adequate leftover risk to the foundation.

STRIDE and DREAD are grounded danger-demonstrating structures for the security evaluation of items and administrations for the duration of their lifecycle (Liu, et al., 2020). For example, STRIDE utilizes information stream graphs for the danger displaying measure. The information stream graphs map framework dangers to the comparing weak framework parts. Given the reliant idea of CPES, an assailant can think twice about framework activity by taking advantage of various part weaknesses. Accordingly, to ensure the general framework security, weaknesses should be tended to both at the part level just as inside the part interrelations (pictured in the information stream outlines). DREAD is considered to be utilized to assess and allot priority to the seriousness of dangers. A DREAD investigation comprises six stages: resource distinguishing proof, framework design arrangement, application deterioration, danger ID, danger documentation, and danger sway rating. DREAD and STRIDE strategies can likewise be utilized together for far-reaching network protection evaluations. Aside from STRIDE and DREAD, different procedures for security evaluations have been proposed and used in the online protection field. OCTAVE, for example, is an optional methodology used by organizations when performing primarily data innovation (IT) security assessments and key getting ready for digital dangers. Nonetheless, late works approve the relevance of OCTAVE for CPS security appraisals, both for the specification of possible dangers just as the plan of countermeasures to keep up with ostensible framework operation. The principle steps continued in OCTAVE security evaluations include the improvement of hazard assessment measures as per functional imperatives, basic resource ID, basic resource weaknesses and relating dangers revelation, and danger sway assessment. DREAD and STRIDE are grounded devices when performing danger displaying examinations and distinguishing weaknesses in the pre-assault context. The examination of enemy conduct post-compromise is likewise significant. Now, the foe has, as of now, conquered the primary line of guard and approaches framework assets. Eminently, there is broad examination on starting double-dealing and utilization of edge guards. In any case, there is an information hole of the foe cycle after starting access has been acquired. To address the previously mentioned entanglement and backing danger displaying, hazard examination, and alleviation procedures, pre- and post-compromise occasions, MITRE fostered the ATT&CK for Enterprise system. MITRE ATT&CK is an open-source information base that incorporates normal antagonistic assault designs (e.g., assaults, methods, and strategies) (Orojloo and Azgomi, 2018). The ATT&CK information base is continually being refreshed with late assault occurrences to upgrade undertaking network protection by uncovering framework weaknesses and warrant more secure functional conditions for organizations and associations. The main part classification incorporates:

1. Resources which comprise control servers, designing workstations, field regulators, human–machine interface (HMI), among others. This load of resources probably will not be evident in each framework. This is figured by the ATT&CK strategy which researches assaults focusing on the individual resources autonomously just as their participation with other modern resources.

2. The second centerpiece is the deliberation zeroing in on the utilitarian limits of the design. Such levels depict the profundity of invasion that the enemy has accomplished. The level reaches from Level 0, which relates to the actual gadgets that arrange the modern cycle, right to Level 2, which incorporates the administrative control frameworks and the designing workstations. The last two pieces of the structure rotate around the antagonistic strategies and methods.

3. The term 'strategies' alludes to the motivation behind why an enemy plays out an activity, i.e., foe objective, for example, upsetting a modern cycle control schedule. Procedures depict the exercises that the foe uses to accomplish the assault objective, i.e., address "how" an assailant achieves his/her targets by making a move, e.g., through adjusting the PLC control rationale.

Researchers working toward robust CPS security and privacy have also developed various risk assessment strategies. We now proceed toward this aspect of risk assessment and state of the art in this area. The expression "hazard appraisal" alludes to the method involved with distinguishing possible dangers and their comparing effect on the framework activity just as deciding systems to alleviate, concede, or acknowledge these dangers dependent on their criticality. Digital danger hazard appraisal is a basic activity that CPES need to perform routinely. The presentation of new advances into CPS alongside the interoperable idea of the upheld ICT foundation builds the dangers emerging from both the digital (e.g., estimation, control orders, or correspondence uprightness assaults) and the actual area. Regularly, hazard evaluation strategies depend on probabilistic investigations that influence Markov-chains, Petri-nets, Bayesian conviction organizations, or game hypothesis to gauge the effect of unfriendly occasions on framework activity. In one more task, for instance, analysts model both the assailants and the framework's safeguards as specialists with various activity sets and targets. Because of the incongruous jobs of such specialists, the relating activity result relies upon the capacity to think twice about the framework's resources or the capacity to identify the vindictive assault according to the point of view of aggressors or on the other hand safeguards, individually. Various works have proposed the worst possible outcome for danger appraisal examinations that utilize extensive Monte Carlo recreations and concentrate on various activity areas of CPS, for example, programmed age control (Kuruvila et al., 2020). Then, at that point, the relationship of such EPS regions with explicit danger relief components is examined by different analysts. For example, the analysts have audited the effect on transports and transmission lines under unusual tasks brought about by digital assaults. They likewise examine how unfriendly situations can be moderated if vigorous assurance framework procedures, for example composed transport and transmission line trippings, are correspondingly set up. Albeit probabilistic danger examinations and most dire outcome imaginable appraisals can give helpful outcomes under explicit requirements (i.e., if by some stroke of good luck some portion of a framework is analyzed), applying such strategies to powerfully evolving enormous scope with the introduction of well-coordinated models can be a difficult undertaking. The huge number of resources extends the pursuit space of thorough techniques, for example, Monte Carlo-based danger analyses. For every resource and each explored possible assault, the danger examination measure should be re-evaluated and recomputed. The danger computation overhead is likewise exacerbated because of the interconnected CPS architecture. The previously mentioned strategies, aside from being computationally serious, can likewise possibly experience the ill effects of poor precision. The security hazard appraisal exactness of these techniques depends on the exact demonstration of the CPS actual parts (Fan, et al., 2019). Failure to appropriately demonstrate CPES can veil interconnection conditions among parts and their layers (digital or physical), and consequently, annoy the danger score estimation measure. The introduced hazard appraisal approaches are trustworthy if evaluation on security is performed somewhat, i.e., they neglect to catch exhaustive framework chances as their emphasis is on explicit pieces of a CPS, disregarding the effect proliferation to the remainder of the foundation. Numerous specialists have proposed procedures that assist the danger appraisal investigation of CPS (since the danger displaying, CPS system examination, and execution measurements assurance have been performed already), and in this way, alleviation strategies can be assessed recursively till the relating risk objectives are met. For instance, if a CPS resource is compromised, there may be different guard systems that could be implemented to relieve the assault. Notwithstanding, the execution of a portion of these systems may bring about critical effects (e.g., uneconomic activity, incomplete framework disengagements, and so on) or influence different pieces of the framework because of its related nature. The capacity to assess, continuously, the adequacy of hazard alleviation instruments gives critical advantages to CPS, meaning to adjust security destinations and framework execution.

The methods of CPS assault can be isolated into two classifications: mass annihilation and accuracy strike. The assault is at first dispatched from web-based media; the age of disinformation messages should be founded on the qualities of the actual framework and the informal organization. In this manner, surveillance is important to recognize, gather, and break down the design and qualities of the actual framework and web-based media. As far as savvy framework disturbance, information gaining might be identified with the interpersonal organization, power framework, coupling between the social and force area, and the organization structure model, the proliferation or falling model. In some cases, aggressors can straightforwardly detect the condition of the CPS by associating with the actual framework and do not have to interrupt the CPS. For instance, when assailants desire to detect the recurrence of force network and afterward increment or abatement the heap to disturb generators, a machine can be utilized straightforwardly to associate with terminal electrical apparatuses to detect the recurrence of the force framework. In the vast majority of the cases, be that as it may, the condition of the actual framework cannot be straightforwardly detected. Assailants need to gather data by interfering into the data framework to assess the condition of the actual framework, which is conceivable. Simultaneously, in light of the fact that assailants just desire to get data about the CPS and do not execute any activity, detecting the CPS state is hard to recognize. Disinformation manufacture is the weaponization of social messages. The objective of these messages is to influence the actual framework precisely; however, its initial step is carried out by controlling the force network clients adequately, i.e., they can be viably spread via online media and perceived by the relating clients. For the most part, the methods of content advancement incorporate twisting realities, trick stories, releasing changed archives, and others. In the wake of making the disinformation, the subsequent stage is the conveyance through online media. The readiness work incorporates:

1. Select framework, foster individuals, and foster the organization: To instigate changes in the heap profiles of individual clients and at last reason significant adjustments in the heap profile of the whole organization, the messages ought to be carefully planned, and maybe, a progression of messages is expected to initiate the clients to approach and understand the aggressor's goal continuously.

2. Publish the substance, infuse content into the media, and dark the beginning: The objective is to spread the disinformation to those clients who will most likely acknowledge it, and their after activity will prompt the normal actual impact.

3. Enhance the proliferation and misdirect people in general: On account of the shrewd network, counterfeit messages might be identified with value data or framework notice, which will initiate the force clients via web-based media to change the electric burdens as the enemies anticipate. One more necessity is the validity of the messages, which guarantees a specific number of clients or explicit clients exercise their true processing ability.

To carry out powerful and effective dispersal, observation and examination of the interpersonal organization are vital, including the dissemination of social clients, their inclination to accept, and coupling with the actual framework (Yang, et al., 2018). It is an enhancement issue to choose the arrangement of social clients whose control could prompt greatest disturbance of the keen framework. Disinformation in various fields has various effects. States can utilize disinformation to practice power over gatherings of individuals. Organizations can utilize disinformation to keep up with or fix their own standing or to harm the standing of a contender. It has been tracked down that the weakness of web-based media and the shrewd network can be taken advantage of, and when the disinformation is diffused true to form, the order and control are going to begin. The assault will enter from the social space, through the digital area, and arrive at the actual space to change the power load. At the point when the distinction between the controlled interest and the genuine interest is enormous, the condition of the actual framework might have an uncommon change.

9.4 Role of Temporal Logic to Develop Threat Models

The main motivation of our chapter is that limited work has been done to develop a framework to address the attack based on the psychological issues about time with which the attacker can formalize the tactics to attempt attacks on the various layers of a CPS ecosystem.

This section will discuss the novel approach for developing threat model by using the robust elements of temporal logic. Temporal logic is being used from the past 50 years to represent the temporal expressions in natural language. There are many reasons why the design of a threat model based on psychology of an attacker will help the CPS designer to be aware of certain elements of weakness and overcome it in the design phase itself. Mitigation techniques become more costly after the attack has occurred and it added to the overall cost of maintaining the security of a CPS. Development of a threat model is an efficient way to minimize this cost. In our work, more importance is given to the way a malicious user thinks while attacking the CPS ecosystem. This thinking of an attacker will not remain the same over the entire duration of the attack. The attacker's strategy may become more aggressive after the attacker gets hold of some resources over the CPS entities. This will leverage the mindset of the attacker to think in alternative attack paths. To model this type of thinking, the attack model must draw support from temporal logic. Nevertheless, first-order temporal logic can help in the development of formal models of threat but they won't help in the development of non-deterministic threat models. Non-deterministic threat models take history of attacks into consideration which may or may not change the state of CPS entity under attack. As the CPS entities can be modeled as a dynamic system rather than static system, variants of temporal logic can be suitable to model the threats by an adversary on the CPS ecosystem.

Threat modeling is a technique used by system or network designers to bring out the weak points in a system before the actual attack happens. After threat modeling is performed, mitigation strategies are developed to counter the threats.

9.4.1 Mathematical Concepts Used in Our Proposed Threat Model

The temporal logic advocates events in two popular models. We use the same models to depict the adversary threat in a highly dynamic system such as the CPS.

1. Time model based on instant
2. Time model based on interval

Time Model based on Instant:
It works on the logic that a backward linear logical model is possible. It means that the past event cannot be changed; it is fixed but future event may be open, meaning multiple branches for future events are possible.

Time Model based on Interval:
Interval-based models for the most part surmise linear time. In any case, they are ontologically more extravagant than moment-based models, as there are a lot more potential connections between time intervals than between time moments.

If we represent time instant as TINS, we can denote the binary relation α of precedence by following equation,

$$TINS = (TINS, \alpha) \tag{9.1}$$

The properties of reflexivity, irreflexivity, transitivity, and asymmetry, anti-symmetry hold true on the above equation.

In a CPS environment, continuous as well as discrete time model events occur. So, threat model should include the concepts of forward discrete and backward discrete models. In forward discrete models, each time instant having a successor will also have corresponding immediate successor. In a backward discrete model, each time instant having a predecessor will also have corresponding immediate predecessor.

We will model the attacker's threat with the help of below three models:

1. Prior's theory of branching time model
2. The Peircean branching time temporal logic model
3. The Ockhamist branching time temporal logic model
    1. Prior's theory of branching time model
       According to this theory, the interrelation between time and modality is denoted by history, which is linear and the present branches into multiple paths. Two instants may share common history. We model the adversary threat ATH as shown below:

$$ATH = (ATH, \alpha) \tag{9.2}$$

The adversary can get hold of sensor data or any other sensitive information and try to fabricate it and use it as input to another CPS device, thereby triggering a malfunction in the whole CPS ecosystem. If the attack fails, then the attacker tries a new path of attack by changing few elements that are part of the old attack. The path of old attack now becomes part of history.

Let us assume, H (ATH) as set of histories spread through a given time instant t in ATH.

For the attacker, $H(ATH)$ contains a sequence of attacks which will eventually harm the system. So, we can now formulate the path of the attack history as given below:

$$H(ATH) = \{A_1, A_2, A_3, \dots A_n\} \tag{9.3}$$

where A_1, A_2, A_3, $\dots$ A_n are different sequence of attacks carried out by the adversary.

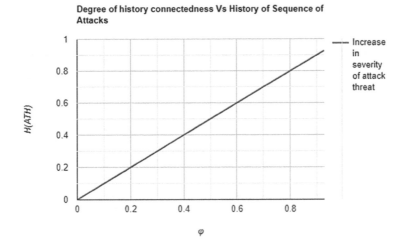

FIGURE 9.1
Plot of degree of history connectedness versus history of sequence of attacks.

Moreover, the attacks $A_1, A_2, A_3, \ldots A_n$ have history connectedness. It means that an attack A_j will be initiated only when an attack A_i prior to it fails to change the correct state of a CPS device into a malicious state. Now, we propose a factor of φ which denotes the degree of history connectedness between the attacks.

So, if the value of $\varphi(A_i, A_j)$ is less, it means A_i and A_j have more dissimilarity in their attack sequence. We assign φ values between 0.00 and 1.00. The above graph shows the relationship between attacks H (ATH) and φ values (Figure 9.1).

We can observe from the plot above that, as φ values increases, the dissimilarity degree between the attacks increases, thereby denoting that the history connectedness between the attacks A_i and A_j are also low.

Now, we discuss the benefits of using this model. This threat model helps us to observe the similarities in the attacks and also helps us to understand the change occurring in the attacks. The attacks following this threat model follow a certain path which can be represented through temporal logic. Nevertheless, the assumptions made while formulating this threat model force us to put efforts in other robust directions.

2. The Peircean branching time temporal logic model
   In this type of temporal logic, we introduce a future operator represented as F_{op}. The contextual semantics of F_{op} is that "it will lead to a case that…". Now, the model can be formulated as,

$$\text{ATH} = (\text{ATH}, F_{op}) \tag{9.4}$$

It infers that now, if $A_1, A_2, A_3, \ldots A_n$ are the various attacks carried out by the adversary on the CPS entity, then future truth is treated as true for all attacks $A_1, A_2, A_3, \ldots A_n$. So, for any two attacks A_i and A_j, there is equal chance for both A_i and A_j to occur. If we want to strengthen the attack model further, then we can include α relation in the attack model. Thus, the attack model can now be represented as,

$$\text{ATH} = (\text{ATH}, \alpha, F_{op}) \tag{9.5}$$

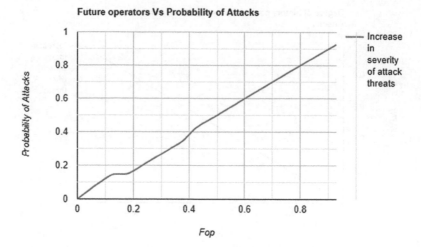

FIGURE 9.2
Plot of future operator versus probability of attacks.

Below a plot between F_{op} and probability of attack is represented for clarity. The probability of attacks is from 0 to 1, and the future operator values are considered from 0.00 to 1.00.

We can observe that as F_{op} values increase, the probability of attacks also increases. This model makes fewer assumptions with respect to history of attacks. Each attack is considered to be equally likely and with same strength of malicious intent. It can be damaging to the sensor data or damaging to the actuator components also in the worst case (Figure 9.2).

3. The Ockhamist branching time temporal logic model
   This model is stronger than the other two models in the sense that it is related to a time instant and also to a history passing through that instant. Therefore, in this model, the future operator F_{op} can be represented as "according to given history, it will result into a situation that...". In this model, in conjunction with F_{op}, we take help of operators P_{st}(Past history of an attack) and operators $\mathcal{E}$, £ which are dual modal operators denoting quantifiers covering the attack history of the adversary.

Subsequently, the attack model can be represented as,

$$\text{ATH} = \text{ATH}(F_{op}, P_{st}, \mathcal{E} \,\&\, \&£) \tag{9.6}$$

The above attack model can be defined inductively. The attack model is interpreted as evaluation of time instant based on backward as well as forward on the given attack history. For P_{st}, there is no branching, so it is definitive and the dual operators $\mathcal{E}$, £ represent alternatives in the future operator. As per the viewpoint from physiological nature of the adversary, we can say that the attacker may act aggressive in future instants of time, if the past history contains failed attacks. This model gives value to the past attack results and how that can influence the attacker to launch attacks in the future. The following graph represents the severity of the attack based on P_{st}, $\mathcal{E}$, £ operators (Figures 9.3 and 9.4). We can find that the percentage of attack severity increases as the value of P_{st} increases. But we

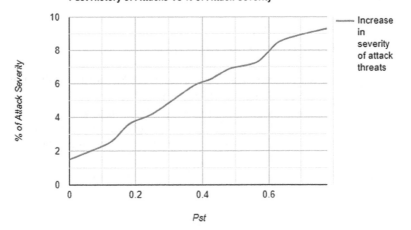

FIGURE 9.3
Past history of attacks versus percentage of attack severity.

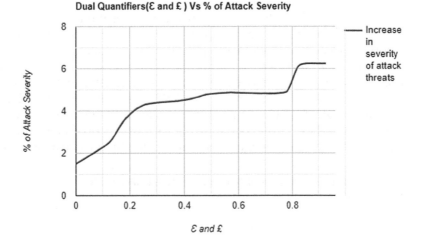

FIGURE 9.4
Dual quantifiers versus percentage of attack severity.

can also observe that the percentage of attack severity does not increase much significantly when compared to function of the quantifiers $\mathcal{E}$, £.

9.5 Experimental Results

We obtained the Kitsune Network Attack dataset from Kaggle, which includes nine attack datasets mined from a network containing many internet of things (IoT) entities. Each dataset has many network packets and various cyber-attacks within it.

Every attack has the following:

1. A preprocessed dataset (csv format)
2. The label vector for each dataset (csv format)

The following graphs show performance of STRIDE, DREAD, and the attack model based on Prior's theory of branching time model (Figures 9.5–9.7).

We can observe that our proposed threat models perform at a better rate in detecting the threats as the count of threats increase. STRIDE and DREAD are popular threat models used in almost 80 to 90% of the CPS ecosystems. Hence, we have picked only these two techniques for performing a comparative analysis.

Our proposed threat models have considered the attacks from the dataset based on the following priority:

Priority 1 (Highest) – People health and safety

Priority 2 (High) – Negative effect in the service or quality of service

Priority 3 (Medium) – Business loss

Priority 4 (Low) – Damage done to the equipment

As the smart world principle is evolving, many medical and health facilities are modifying themselves to become a part of the smart revolution, which makes security of CPS entities even more important. Hence, we have developed the threat models suitable to apply on e-health systems primarily. Next, we have considered the attack model which may affect the quality of services and also the service may be made unavailable by the adversary. This causes imbalance in the overall functionality of the CPS ecosystem. So, we have given high priority to such cases of attack. The next priority is assigned to the financial loss occurred due to the attack which can be controlled by applying various security frameworks. The least importance is given to equipment damage as they are replaceable after an attack is identified.

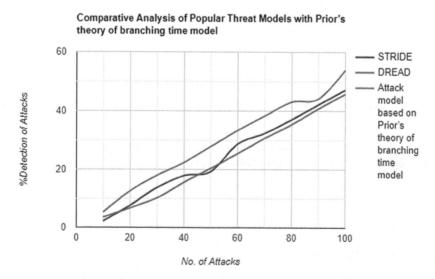

FIGURE 9.5
Performance of popular threat models against attack model based on prior's theory of branching time.

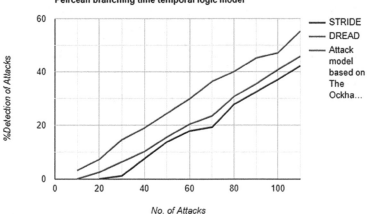

FIGURE 9.6
Performance of popular threat models against attack model based on the peircean branching time temporal logic.

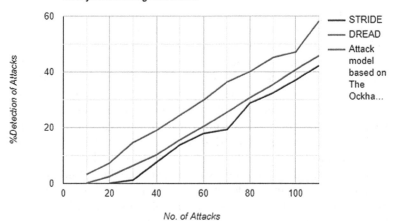

FIGURE 9.7
Performance of popular threat models against attack model based on the ockhamist branching time temporal logic.

9.6 Future Directions

Our threat models cannot be applied to certain attacks. This is due to the reason that CPS devices are heterogeneous. All devices do not behave in the same manner, so we cannot ascertain at which instant of time which behavior of the device may be used by the adversary to attack. Also, depending on design constraints, choosing appropriate design model may be difficult than it seems. Another challenge is to store the specifications from a given instant from attack history. We have tabulated the various factors and their degree of suitability for the proposed threat models (Table 9.1).

TABLE 9.1

CPS Attack Factors and Their Degree of Suitability for Proposed Threat Models

Threat Model	Threat Factors (Degree of Suitability)					
	Source/Attacker	Target	Motive	Attack Vector	Availability	Denial-of-Service
Prior's theory of branching time model	High	High	High	High	High	Low
The Peircean branching time temporal logic model	High	High	High	High	Medium	Low
The Ockhamist branching time temporal logic model	High	Low	High	High	High	Low

We can easily observe that the Prior's theory of branching time model is having low suitability for DoS attacks because the attackers action does not have any co-correlation with history of sequence of attacks. Next, we can infer that the Peircean branching time temporal logic model is also having low suitability to DoS attacks and medium suitability to the availability attacks. This is because an attack that affects the service of a CPS node makes it stop the availability of data which has to be given as input to a sensor node. The Peircean branching time temporal logic model is not guaranteed to handle such attacks at a 100% successful rate. Finally, we can observe that the Ockhamist branching time temporal logic model is not so much suitable to targets and DoS attacks. The target may be in this case a CPS entity like a sensor or an actuator which are generally suffering from power failure attacks and physical attacks. The Ockhamist branching time temporal logic model does not stop the adversary from launching physical attacks where the components are damaged so only estimation of such threats can be modeled by the CPS security designer.

In future, security designers of CPS need to develop threat models based on multiple quantifiers based on temporal logic. The adversary may use hybrid mechanisms for attacks which make our threat model weak as there are no provisions of how hybrid attacks that can be handled by our proposed threat models. Another aspect to be considered vigorously is the confidence with which the models are developed on high-risk CPS entities. Few CPS ecosystems are so much time-sensitive that recording the different states in their history may be difficult. Even if the threat model designers are able to record the history, the confidence level to use such history may not be high. This remains quite a challenge to be solved in future. The final challenge is the models may suffer from some history perturbations, which may disturb the attack correlations, thereby making the threat models ineffective. This loss may cause significant reduction in the degree of suitability of an attack model with respect to the attack vector. So, the challenge for the designers in future is that history perturbations occurring in the history are isolated or their effect is nullified by the quantifiers used in the threat model.

9.7 Conclusion

This chapter brings out the application threat models using temporal logic. We have specifically used the Prior's theory of branching time model, the Peircean branching time temporal logic model, and the Ockhamist branching time temporal logic model. These models have added strength to the development of CPS threat models by adding the element of dynamic and adverse thinking of a CPS attacker. The experimental results show that the performance of temporal models outperforms the traditional threat models of STRIDE and DREAD, considering standard datasets used to validate the popular datasets. The future directions indicate further improvements are possible in the proposed threat models due to the inherent limitations arising from the assumptions taken while developing the application logic of the threat models and mapping them into the scope of temporal logic. Overall, through this chapter, we strongly advocate the suitability of the threat models as they are developed on the psychology of the adversaries combined with provisions to accommodate uncertainties of time.

References

D. Gritzalis, M. Theocharidou, and G. Stergiopoulos, *Critical infrastructure security and resilience*, Cham: Springer, January, 2019, pp. 0–16.

A. Keliris, C. Konstantinou, N. G. Tsoutsos, R. Baiad, and M. Maniatakos, "Enabling multi-layer cyber-security assessment of industrial control systems through hardware-in-the-loop test-beds," in *2016 21st Asia and South Pacific Design Automation Conference (ASP-DAC)*. IEEE, 2016, pp. 511–518.

C. Konstantinou, "Towards a secure and resilient all-renewable energy grid for smart cities," *IEEE Consumer Electronics Magazine*, 11, 33–41, 2021.

S. Muyeen and S. Rahman, *Communication, control and security challenges for the smart grid*. The Institution of Engineering and Technology, 2017.

J. Ospina, X. Liu, C. Konstantinou, and Y. Dvorkin, "On the feasibility of load-changing attacks in power systems during the covid-19 pandemic," *IEEE Access*, 9, 2545–2563, 2021.

B. Chen, K. L. Butler-Purry, A. Goulart, and D. Kundur, "Implementing a real-time cyber-physical system test bed in rtds and opnet," in *2014 North American Power Symposium (NAPS)*, 2014, pp. 1–6.

C. Queiroz, A. Mahmood, and Z. Tari, "Scadasim—A framework for building scada simulations," *IEEE Transactions on Smart Grid*, 2(4), 589–597, 2011.

N. Dorsch, F. Kurtz, H. Georg, C. Hägerling, and C. Wietfeld, "Software-defined networking for smart grid communications: Applications, challenges and advantages," in *2014 IEEE International Conference on Smart Grid Communications (SmartGridComm)*. IEEE, 2014, pp. 422–427.

H. Georg, S. Müller, N. Dorsch, C. Rehtanz, and C. Wietfeld, "Inspire: Integrated co-simulation of power and ICT systems for real-time evaluation," in *2013 IEEE International Conference on Smart Grid Communications (SmartGridComm)*, 10, 2013, pp. 576–581.

M. J. Stanovich, I. Leonard, K. Sanjeev, M. Steurer, T. P. Roth, S. Jackson, and M. Bruce, "Development of a smart-grid cyberphysical systems testbed," in *2013 IEEE PES Innovative Smart Grid Technologies Conference (ISGT)*, 2013, pp. 1–6.

S. Boschert and R. Rosen, *Digital Twin—The Simulation Aspect*. Cham: Springer International Publishing, 2016, pp. 59–74. [Online]. Available: https://doi.org/10.1007/978-3-319-32156-15

M. Zhou, J. Yan, and D. Feng, "Digital twin framework and its application to power grid online analysis," *CSEE Journal of Power and Energy Systems*, 5(3), 391–398, 2019.

T. Huang, F. R. Yu, C. Zhang, J. Liu, J. Zhang, and Y. Liu, "A survey on large-scale software defined networking (SDN) testbeds: Approaches and challenges," *IEEE Communications Surveys & Tutorials*, 19(2), 891–917, 2016.

J. Ospina, N. Gupta, A. Newaz, M. Harper, M. O. Faruque, E. G. Collins, R. Meeker, and G. Lofman, "Sampling-based model predictive control of PV-integrated energy storage system consider-ing power generation forecast and real-time price," *IEEE Power and Energy Technology Systems Journal*, 6(4), 195–207, 2019.

A. Keliris, C. Konstantinou, M. Sazos, and M. Maniatakos, "Lowbudget energy sector cyberattacks via open source exploitation," in *2018 IFIP/IEEE International Conference on Very Large Scale Integration (VLSI-SoC)*. IEEE, 2018, pp. 101–106.

Y. Xiang, L. Wang, and N. Liu, "Coordinated attacks on electric power systems in a cyber-physical environment," *Electric Power Systems Research*, 149, 156–168, 2017.

J. Tian, B. Wang, T. Li, F. Shang, and K. Cao, "Coordinated cyberphysical attacks considering dos attacks in power systems," *International Journal of Robust and Nonlinear Control*, 30(11), 4345–4358, 2020.

H. Tu, Y. Xia, K. T. Chi, and X. Chen, "A hybrid cyber attack model for cyber-physical power systems," *IEEE Access*, 8, 114 876–114 883, 2020.

K. Pan, A. Teixeira, M. Cvetkovic, and P. Palensky, "Data attacks on power system state estimation: Limited adversarial knowledge vs. limited attack resources," in *IECON 2017—43rd Annual Conference of the IEEE Industrial Electronics Society*, 2017, pp. 4313–4318.

A. Barua and M. A. Al Faruque, "Hall spoofing: A non-invasive dos attack on grid-tied solar inverter," in *29th flUSENIXg Security Symposium (flUSENIXg Security 20)*, 2020, pp. 1273–1290.

V. Venkataramanan, A. Hahn, and A. Srivastava, "Cp-sam: Cyberphysical security assessment metric for monitoring microgrid resiliency," *IEEE Transactions on Smart Grid*, 11(2), 1055–1065, 2019.

Y. Zhang, V. Krishnan, J. Pi, K. Kaur, A. Srivastava, A. Hahn, and S. Suresh, "Cyber physical security analytics for transactive energy systems," *IEEE Transactions on Smart Grid*, 11(2), 931–941, 2019.

A. P. Kuruvila, I. Zografopoulos, K. Basu, and C. Konstantinou, "Hardware-assisted detection of firmware attacks in inverter-based cyberphysical microgrids," arXiv preprint arXiv: 2009.07691, 2020.

X. Liu, J. Ospina, and C. Konstantinou, "Deep reinforcement learning for cybersecurity assessment of wind integrated power systems," *IEEE Access*, 8, 208 378–208 394, 2020.

H. Orojloo and M. A. Azgomi, "A stochastic game model for evaluating the impacts of security attacks against cyber-physical systems," *Journal of Network and Systems Management*, 26(4), 929–965, 2018.

Y. Fan, J. Li, D. Zhang, J. Pi, J. Song, and G. Zhao, "Supporting sustainable maintenance of substations under cyber-threats: An evaluation method of cybersecurity risk for power cps," *Sustainability*, 11(4), 982, 2019.

Y. Yang, S. Wang, M. Wen, and W. Xu, "Reliability modeling and evaluation of cyber-physical system (CPS) considering communication failures," *Journal of the Franklin Institute*, 358, 1–16, 2018.

10

Software-Defined Network Security

Omerah Yousuf and Roohie Naaz Mir

National Institute of Technology, Srinagar, India

CONTENTS

DOI: 10.1201/9781003241348-10

10.1 Introduction

SDN (software-defined networking) approach to network design has been progressively gaining traction. As a result, an increasing number of businesses are planning their network architecture with this strategy in mind. Traditional networking necessitates the purchase of additional equipment to expand network capacity. The difference between SDN and conventional networking is that the former requires additional equipment to expand, whereas the latter only requires inputs. In a typical computer network, routing is handled by routers, switches, and gateways, which are connected via connections and nodes. These are the network elements that apply quality of service (QoS) factors to decide. These network components are unique to each provider. As a result, these components' programmability is limited to the network vendor's standards and compatibility (Fraser et al., 2013). To remove this limitation and make networks more programmable, SDN was developed. The control plane and the data plane are the two planes that make up the network. The control plane is the network's brain, and it is from here that the entire network may be managed to fit the application's demands, regardless of vendor standards. When data are requested, the data plane is a forwarding aircraft that delivers it to the control plane (Prathima Mabel, Vani, & Rama Mohan Babu, 2019).

SDN is slowly becoming a reality, with a slew of SDN-enabled devices in the works. Cloud computing and virtualization technologies, which have long been explored in the research sector, have found commercial application in the merging of discrete control and data-plane tasks, as well as network programmability (Scott-Hayward, O'Callaghan, & Sezer, 2013). Because SDN networks are expected to hold private and confidential information at times, they must comply with the same security standards as traditional networks. SDN significantly changes the design and inter-communicative properties of network components, creating an altogether new platform for attackers attempting to compromise security. As a result, similar levels of security as for traditional networks are required, but to guard against a wider spectrum of assaults (Spooner & Zhu, 2016). SDN, while allowing for effective network management, has been discovered to be more vulnerable to attacks than traditional networks, with attacks on forwarding devices and controllers affecting network communication principles, making the network vulnerable to security attacks and issues.

Chapter organization: The rest of the chapter is organized as follows: A quick summary of SDN development and basic principles is presented in Sections 10.1 and 10.2. The security analysis of SDN is discussed in Section 10.3. Section 10.4 discusses the many issues that have arisen throughout the development of the SDN. The various issues and solutions in SDN architecture are described in Sections 10.5 and 10.6. SDN in CPS is briefly discussed in Section 10.9. Finally, Section 10.8 wraps up the chapter and discusses future study objectives. Finally, the references are listed.

10.1.1 SDN Evolution

Stanford University was the first to establish the notion of SDN in 2005. SDN's main goal was to make networks easier to create by separating network logic from forwarding hardware. SDN has been around for more than a decade. Many people thought that dazzling software-defined networks would eventually supersede tightly coupled, vertically integrated network solutions when SDN was originally announced. The three stages of SDN's history are depicted in Figure 10.1. Each stage makes a unique contribution to history (Feamster, Rexford, & Zegura, 2014; Stancu, Halunga, Suciu, & Vulpe, 2015):

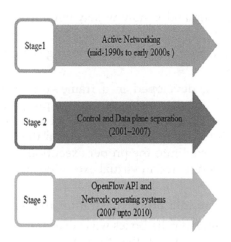

FIGURE 10.1
SDN evolution.

i. Active networks (mid-1990s to early 2000s), which allowed for further innovation by introducing programmable functions to the network;

ii. Separation of the control and data planes (between 2001–2007), resulting in open control and data plane interfaces;

iii. The OpenFlow application programming interface (API) and network operating systems were the first major open interface applications, and they researched ways for making control-data-plane separation scalable and believable from 2007 until roughly 2010.

Stage 1: Active networking

Active networking was a groundbreaking concept because it created a programming interface that exposed resources on individual network nodes and allowed for the building of tailored functionality that could be applied to a portion of packets flowing through the node. One of the first "blank slate" approaches to network architecture was active networking. There were two programming paradigms in active networking:

- Capsule model – data packets carrying code to be executed at nodes were carried in-band.

- Programmable router/switch model – Out-of-band techniques transmitted the code to be executed at the nodes.

Despite the fact that active networks were not widely used, the three significant contributions to SDN were:

i. **Network functionalities that can be programmed to decrease the barrier to innovation**

Active networks were among the first to propose the use of programmable networks to address the sluggish pace of computer networking progress. Although many early SDN concepts emphasized increasing the programmability of the control plane, active networks promoted programmability of the data plane. Recent SDN research is looking on the growth of SDN protocols like OpenFlow to provide a broader variety of data-plane tasks. Furthermore, the notion of segregating experimental traffic from regular traffic originated in active networking and is now widely employed in OpenFlow and other SDN technologies.

ii. **Network virtualization and de-multiplexing software programs based on packet headers**

Active networking developed in a framework that specified a platform that would allow for the testing of various programming paradigms. The demand for network monitoring arose as a result of this requirement. This paradigm was interestingly taken forward in the Planet-Lab design, packets are de-multiplexed into the proper execution environment, and various experiments are conducted in virtual execution environments depending on the packet headers.

iii. **Idea of a unified middle-box orchestration architecture**

Unified control over middle-boxes was never fully achieved in the period of active networking. Although the study had no direct influence on network function virtualization (NFV), some of the findings may now be applied to unified architecture deployment.

Stage 2: Control and data-plane separation

During this period, traffic quantities steadily increased, making network dependability, reliability, and performance more significant. Better network management features, such as control over traffic delivery channels, were sought by network operators (traffic engineering). They noticed that because of the tight integration of the control and data planes, classic routers and switches had a difficulty with network administration. Attempts were made to separate the two after this finding. Although the first attempts to divide the control and data planes were pragmatic, they constituted a considerable conceptual shift from the Internet's traditional tight coupling of route calculation and packet forwarding. The following are the concepts that were explored in SDN when segregating the data and control planes:

✓ Centralized access control with an open data-plane interface

✓ State management on a large scale

Stage 3: OpenFlow API and network operating systems

Researchers and funding organizations were interested in the notion of large-scale network testing; hence, OpenFlow was established. It managed to strike a compromise between the ambition of completely programmable networks and the realities of assuring their implementation in the actual world. Nevertheless, OpenFlow included many of the concepts from previous work on the separation of the planes; it would still be a work in progress.

10.2 Software-Defined Networking Concepts

10.2.1 Definition

SDN is still in the process of developing an operational layer of abstraction. There are several general definitions of SDN, each of which is related to the degree of dispersed control plane maintained. The Open Networking Foundation (ONF) is a non-profit organization committed to development, certification, and commercialization of SDN. The ONF's most thorough and commonly recognized definition of SDN is as follows (Xia, Wen, Foh, Niyato, & Xie, 2015):

SDN is a network architecture that decouples network control from forwarding and makes it programmable directly. In contrast to traditional network architecture, SDN is a paradigm change that uses software-based controls to make policy execution and implementation easier. It uses a well-defined application programming paradigm to abstract data and manage the actions of networking devices such as routers and switches.

10.2.2 OpenFlow for SDN

The ONF created OpenFlow as a multivendor standard for deploying SDN in networking equipment. The OpenFlow protocol is used to connect an OpenFlow controller and an OpenFlow switch, as shown in Figure 10.2. The OpenFlow switch may receive instructions from the OpenFlow controller on how to handle incoming OpenFlow data packets (Vaughan-Nichols, 2011).

The OpenFlow switch can be configured to: (1) identify and recognize packets from an input port based on various packet header fields; (2) handle the packets in various ways, and (3) reject or progress the packets to a designated outgoing port.

OpenFlow commands are sent from an OpenFlow controller to an OpenFlow switch to create "flows". Each flow contains packet match fields, flow precedence, unique counts, packet-processing instructions, flow timeouts, and a cookie. Tables are used to organize the flows. Before it departs on an egress port, an incoming packet OpenFlow is an open protocol that may be used to control the flow table. Traffic can be divided into production and research flows by a network administrator. Researchers may choose which paths their packets go and how they are processed, allowing them to manage their own flows. Over the same network, operational traffic is separated and routed in the same manner it does today (Hu, Hao, & Bao, 2014; Upc, 2012).

10.2.3 Architectural Model

The SDN architecture defines how the SDN functions at various levels and how it ensures software security and dependability. In SDN, there are three layers: data, control, and

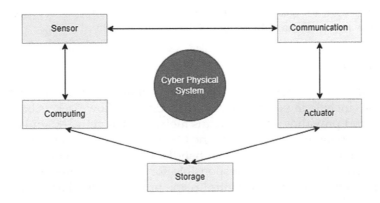

FIGURE 10.2
OpenFlow switch.

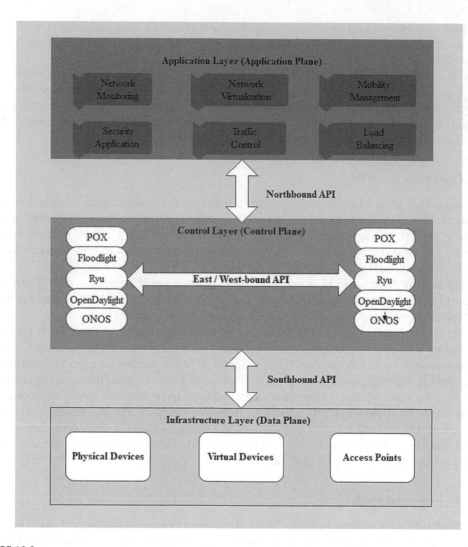

FIGURE 10.3
SDN architecture.

application as depicted in Figure 10.3. Northbound and southbound APIs are used to interact between these levels (Rana, Dhondiyal, & Chamoli, 2019).

10.2.3.1 Data Plane

This plane is made up of forwarding devices that make up network architecture and whose key duties include imposing forwarding actions on flow packets based on controller instructions and providing network status metrics when network applications request them. It is similar to the physical layer of the OSI model and is made up of network elements such as real and virtual devices that handle data flow. It is known as the SDN forwarding plane, and it is physically responsible for passing frames of packets from its entrance to egress interface utilizing control-plane protocols (Pradhan & Mathew, 2020).

10.2.3.2 Control Plane

This plane houses the controller, which is in charge of flow control, routing, error control, transmission control, security, and other network-critical tasks. The controller is the network's brain, housing all of the network's intelligence. It has a programmable interface for establishing network QoS settings, as well as routing, flow management, error control mechanisms, and security policies. It is a logical entity that accepts application-layer tasks and sends them to SDN components. The controller's function is to gather important data from hardware devices and relay it to SDN applications in the form of an abstract network picture that comprises various network events (Nunes, Mendonca, Nguyen, Obraczka, & Turletti, 2014).

10.2.3.3 Application Plane

The application plane is a layer that contains applications and services that initiate network function requests from the Control Plane and Data Plane. Different applications that are used in a business to inform the network what to do based on the business's needs, controllers employ APIs to deliver commands to routers, SDN switches, and other devices to complete the needed task.

Northbound and southbound interfaces, often known as SDN architectural APIs, define communication between applications, controllers, and networking systems. The controller communicates to applications through a northbound interface, whereas the controller links to actual networking devices via a southbound interface. Because SDN is a virtual network overlay, these components do not need to be physically located in the same region.

10.3 Security Analysis of SDN

SDN facilitates communication network innovation by allowing the network to be programmable and consolidates network control for improved resource visibility, simplified network administration, consistent network policy enforcement, and simplicity of new function deployment, among other benefits. However, if these qualities are ignored, they can render SDN to extremely vulnerable security concerns. Indeed, security is a significant difficulty, and ONF has established a security working group to address this issue in SDN. As a result, it is vital to look at the security problems that SDNs encounter and come up with solutions to them.

10.3.1 SDN for Improving Security

Enterprise networks today have the issue of being constantly attacked by new threats. To meet this issue, new business network management and protection technologies are required to enhance security. This necessitates modifications in network architecture and security. SDNs are better positioned to adapt to these difficulties at this level of network growth.

As the network edge fades, internet of things (IoT) will become a greater security issue. The threat landscape for networks is becoming more dangerous as a result of millions of insecure IoT devices. Bad actors may exploit vulnerabilities in IoT devices to conduct

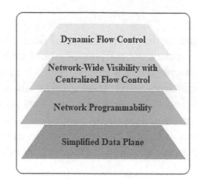

FIGURE 10.4
Security features of SDN.

distributed denial-of-service (DDoS) attack and other attacks as IoT devices become more widely used. In a world of more complex attack vectors, enterprise network security must be the line in the sand for securing data and assets. Understanding how to utilize SDN architecture to create a more flexible and adaptive network security foundation is the greatest method to fortify that line of defense. SDNs design has a number of strong security features shown in Figure 10.4 and discussed as follows (Seungwon Shin, Xu, Hong, & Gu, 2016):

A. **Dynamic flow control**

Because of the basic qualities of SDN, a network application may dynamically govern network flows. In terms of network security, controlling network flows dynamically offers several benefits, including: (i) with SDN, we can more efficiently manage network flows without the need for additional middleware. We may also utilize an OpenFlow switch/router or any network device with SDN capabilities for access control; (ii) it also makes distinguishing between harmful (or suspect) network traffic and benign network data in real time much easier.

B. **Flow control from a centralized location with network-wide visibility**

"Network-wide visibility" refers to the ability for network control to see the status of any network device installed anywhere at any time. SDN can assist with network-wide monitoring as well as identifying and defending against network-wide assaults. The SDN can easily monitor any network device by collecting network data and receiving flow request signals from them because of its fundamental properties (i.e., control and monitor full networks in a centralized manner). This feature may help to enhance the use of security equipment by directing defined network flows to necessary/specific security appliances, such as hardware components, middle-boxes, and virtual network functions of networking components.

C. **Network programmability**

SDN allows network behavior to be controlled independently of the networking hardware that provides physical connectivity. As a result, network operators may alter their networks' functioning to enable new services and even specific clients. A network operator can, for example, parameterize the data plane depending on the components that make up the network's environment, thanks to network programmability, which supports network flexibility. It enables the deployment of a wide range of security services and systems as network apps, either as controller applications or in their own security applications layer.

D. **Simplified data plane**

Because instructions are given by SDN controllers rather than numerous, vendor-specific devices and protocols, SDN simplifies network design and operation when done using open standards. The data plane has relatively simple logic because the data plane and control plane are separated in the SDN design. This streamlined data plane allows us to experiment with new features. This idea may be extended to security as well. SDN's data plane may be extended to make it more appropriate for security.

10.3.2 Need of SDN Security

Without adding SDN security challenges to the mix, traditional network security vulnerabilities are problematic enough. However, by using SDN, enterprises potentially expose their networks to new forms of vulnerabilities and assaults, especially if they do not have sufficient strategies in place. The SDN controller is a major source of security issue with SDN. The network's intelligence is housed in the controller. The controller is in the hands of everyone who has access to it. This implies that businesses must set up rules and network architecture to ensure that the correct people are in charge. Because successful assaults on the SDN controller can completely interrupt network operations, it is an important aspect of the security conversation. Organizations may use role-based authentication to prevent these threats by ensuring that only the appropriate personnel have access to systems and data. Organizations who want to keep on top of SDN security risks should focus on following three key areas (Hayajneh, Bhuiyan, & McAndrew, 2020):

- Data privacy must be protected.
- Safeguard the system's reliability.
- Make it sure, that the network's services are available as and when needed.

Network features must be preserved in SDN-based networks. Although the network's programmability is enhanced by the open interface and centralization, it also exposes the network to assaults and unauthorized access at multiple points. The controller is the key network node in this arrangement, making it vulnerable to assault. Exploiting a poorly secured controller, an attacker can get access to the whole network. Attackers can also introduce fraudulent flows and impair the network's operation by compromising the data plane or the secure link between the data plane and the control plane. An attacker can simply introduce undesired flows to flow tables and reroute data to unexpected destinations by using security compromise switches. This might result in denial-of-service (DoS) attacks, man-in-the-middle attacks, eavesdropping, and other issues. As a result, developing effective ways for preventing and safeguarding the controller against assaults is critical (Dargahi, Caponi, Ambrosin, Bianchi, & Conti, 2017).

10.4 SDN Challenges

SDN provides a new type of minimal hardware for software development communities to rally around in order to provide cost-effective, reliable network services suited to individual demands by separating the data and control planes of traditional networks. Advanced

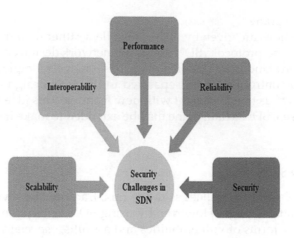

FIGURE 10.5
Challenges in SDN.

businesses and networking organizations, on the other hand, must overcome a variety of challenges in order to reap the full benefits of SDN. The various challenges in SDN associated with each tier of the framework, as well as the interactions between these layers, are as follows (Deshpande, 2013) and are depicted in Figure 10.5.

10.4.1 Scalability

Scalability refers to a system's, network's, or process's ability to handle rising quantities of work while also expanding to meet that growth. As a result, a centralized controller, or a cluster of controllers, may efficiently handle data forwarding node control-plane services for a large number of nodes. To improve scalability, the theoretically consolidated controller must be physically distributed out rather than functioning peer-to-peer. Whether the controller design is distributed or peer-to-peer, network devices will suffer the same challenges as the controller when contact occurs. Because SDN contains centralized or distributed controllers that connect with data planes on separate systems, the controllers may operate as a network bottleneck. Large networks, in particular, can overwhelm controllers because of the enormous amount of networking requests. The bottleneck tightens as networks grow, and network performance worsens as a result.

A decentralized control architecture or a comparable solution, such as segmented or entirely distributed control planes, can help to increase scalability. Such methods, however, may bring additional challenges, such as attaining convergence and managing a large number of control instances.

10.4.2 Interoperability

All network devices are SDN-ready, making SDN straightforward to implement in new networks. Converting to SDN presents a unique set of challenges because the legacy network is likely to include active business and networking components. Most businesses and networks will need to shift to SDN, which will demand a period of compatibility with hybrid legacy-SDN architecture. SDN and traditional network nodes can work together if a protocol is developed that supports SDN communications while maintaining backward compatibility with existing control-plane technologies. When switching to SDN, the cost, risk, and disruption of services are all minimized.

10.4.3 Performance

Performance is the most serious issue faced by all networks. A network's durability, security, scalability, or interoperability are all irrelevant if it does not function well. It may face delays because to the unique control and data plan configuration. This can result in an unsustainable degree of latency in big networks, lowering network performance. Furthermore, when controller reaction time and throughput are combined, poor performance might occur, complicating scaling.

Many performance concerns in vast and dynamic networks can be handled by putting more intelligence in the data plane or switching to a distributed control-plane design. While this improves SDN, it also replicates conventional networks based on widely dispersed intelligent devices, which is at odds with SDN's initial objective. Virtualization must be maintained without jeopardizing network performance or establishing single points of failure, necessitating a delicate balancing act.

10.4.4 Reliability

While developing any software, reliability is critical; if a system fails, users should be notified, and the remedy should work automatically. Software reliability refers to the likelihood that it will perform as expected in a given environment and for a certain period of time. The network management setup of the SDN controller must be intelligent and validated in order to boost network availability and avoid and handle problems. To preserve flow control and continuity in legacy networks, network traffic is redirected through other adjacent paths/nodes when devices fail or cease operating.

10.4.5 Security

Up to this point, security vulnerabilities connected with SDN have been the topic of research. Security issues must be considered when SDN becomes more widely utilized and deployed. The ONF organized a security working group with them with this in mind. Authentication and authorization at the controller-application level are considered to be among the most dangerous security flaws. SDN security must be incorporated into the architecture and made available as a service to ensure the trustworthiness, authenticity, and secrecy of all associated resources and information. In order to offer security inside the architecture, the user must do the following tasks.

- Securing controller: Must strictly control the access to the SDN controller.
- Protecting the controller: If the network goes down (for example, as a result of a DDoS attack), the SDN controller goes down with it; therefore, its accessibility must be maintained.
- Establish trust: It is essential to keep the network's communications secure. This involves making sure that the SN controller, applications, and the devices it manages are all functional and dependable.
- Establish a firm policy framework: A frequent system review is essential to ensure that the SDN controllers are performing exactly what you want.

- Execute scientific investigations and remediation: You must be able to find out what went wrong, recover from it, most likely condemn it, and then guard yourself in the future if something happens wrong.

10.5 Security Issues to the SDN Architecture

Separating the planes and combining control-plane tasks into a single system may be necessary for future networks, but it also poses new security vulnerabilities. Communication connections between planes, for example, can be targeted to conceal one plane so that the other can be attacked. Because of its great visibility, the control plane is more exposed to security threats, particularly DoS and DDoS assaults. Figure 10.6 depicts the numerous dangers and assaults that occur at various tiers of the SDN architecture and is explained as follows (Ahmad, Namal, Ylianttila, & Gurtov, 2015), (Correa Chica, Imbachi, & Botero Vega, 2020):

- **Application-layer issues**

  The application plane on a network is the layer of applications and services that integrates network capabilities from both the control and data planes. This tier contains traditional network devices. The functionality of these devices is essentially the same in SDN, but transmission is encapsulated, consolidated, and usually outsourced. This layer is susceptible to the following security flaws:

  i. **Authentication and authorization**

     The lack of an authentication technique between SDN controllers and network applications is one of the most critical security issues with the SDN paradigm. The majority of the control plane's functions are carried out by a range of applications that run on top of the controller and are often developed by third-party

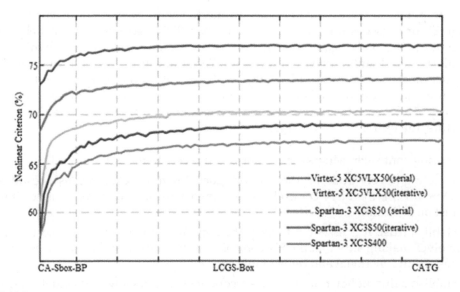

FIGURE 10.6
Various security issues in SDN.

organizations or engineers rather than the controller's manufacturer. These programs can gain access to network resources if there are no enough security safeguards in place. As a result, third-party application authentication and permission constitute a severe security risk in SDN networks with centralized (logical) controllers (Wen, Chen, Hu, Shi, & Wang, 2013).

ii. **Access control and accountability**

Business logic is used by applications in SDN to ease network connection. To make the SDN network safe, proper access control and accountability procedures are essential. Trust may be leveraged to execute a number of assaults throughout the whole network after a hostile or compromised program has gotten access to the controller. An attacker can get access to and manage the whole network's resources if he impersonates an application. Applications in SDN are divided into two categories: (i) SDN-aware applications directly locate and interact with SDN controllers; (ii) SDN-unaware applications communicate via application datagrams in particular forms. In the second scenario, an infiltrated real SDN application can be exploited as a gateway to get unwanted access to the network control plane. Another challenge is maintaining accountability for layered apps' usage of network resources (Foschini, Mignardi, Montanari, Internet, & 2021, 2021).

- **Control-plane issues**

In the SDN paradigm, where decisions are made in a centralized way on the control plane, the controller is crucial. As a result, it is a prime target for nefarious activity. The controller's security is particularly important in terms of its capacity to properly authenticate and segregate apps. The controller thus is facing many security issues itself which are discussed as follows:

i. **Scalability-related hazards**

The controller, which makes logically centralized forwarding choices, handles the majority of OpenFlow's complexity. If the controller is required to apply flow rules for each new data stream flow, it might quickly become a bottleneck. When employing OpenFlow on high-speed networks with 10 Gbps lines, it was observed that today's controller implementations are unable to handle the huge quantity of new flows. The lack of scalability, according to Seungwon Shin, Yegneswaran, Porras, & Gu (2013), allows targeted assaults to induce control-plane saturation, which has harsher consequences in SDNs than in conventional networks. Because of its scalability, the controller is a common target for DoS and distributed DoS assaults.

Another challenge for existing controller implementations is determining the number of forwarding devices that must be maintained by a single controller in order to meet the delay restrictions. The time period will very certainly grow in lockstep with the quantity of flows on the controller, which is mostly determined by its processing capability. A single point of failure might result from this limitation in controller capabilities (Yao, Bi, & Guo, 2013).

ii. **DoS attacks**

The most serious security threats to the SDN controller are DoS and distributed DoS attacks. If the SDN controller is hacked or rendered unreachable for switch requests, the entire network will come to a halt. As the number of flows in the data line grows, the switches will overload the controller with flow setup requests, causing it to perform badly.

DDoS is a well-known assault against the centralized controller, in which a large number of packets with non-repetitive headers are injected into the network from a group of compromised (zombies) servers using spoofing and other means. The burden of these packets eventually gets so great that the controller is no longer accessible for normal network operations (Ali et al., 2020).

iii. **Distributed control-plane challenges**

To accommodate a vast number and diversity of devices that cannot be managed by a single SDN controller, many controllers must be deployed, separating the network into various subdomains. Information aggregation and preserving unique privacy standards in each sub-network will be difficult if the network is fragmented into numerous software-defined sub-networks (Bannour, Souihi, & Mellouk, 2018).

iv. **Application-based issues**

The control plane's security may be jeopardized by applications that operate on top of it. In general, controller security presents a difficulty in terms of the controller's capacity to authenticate services and approve resources utilized by applications while maintaining sufficient isolation, auditing, and monitoring. Furthermore, before access to network information and resources is granted, certain applications must be segregated depending on their security implications. Separating applications is necessary for explicitly authenticating and approving third-party and operator apps, as well as setting audits for each program (Mubarakali & Alqahtani, 2019).

- **Data-plane issues**

The SDN data plane caches the flows following the packet's transmission to the controller for the purpose of setting up flow rules. The data-plane security issues of OpenFlow flows are being investigated. Furthermore, with SDN, the forwarding devices are simplified and heavily reliant on the controller. As a result, the forwarding devices are unable to distinguish between legitimate and awed or malicious flow rules. Because of the buffer flow constraint, the data plane is vulnerable to many forms of assaults because malicious flows are stored in the buffer. The various security issues at this plane are discussed as follows:

i. **Flow rules installation**

The OpenFlow controller adds flow rules into the flow tables of OpenFlow switches in OpenFlow networks. These flow rules can be configured either before or after a new host begins delivering packets (proactive rule installation) or after the first packet is transmitted (reactive rule installation) (reactive rule installation). Because switches no longer have decision-making authority, recognizing valid flow rules and differentiating them from fraudulent or hostile ones is the first and most important security challenge.

ii. **Injection of a faulty control packet**

When a target switch is vulnerable to fuzzing, it can be driven into an unwanted state by receiving forged control packets with incorrect or misused headers that are ingeniously engineered to reveal existing vulnerabilities or erroneous behavior seen under inappropriate input conditions (Yoon et al., 2017).

iii. **Switch-controller link**

The data plane in split architectures such as SDN stops operating if a switch fails to get forwarding instructions from the control plane because of control-plane

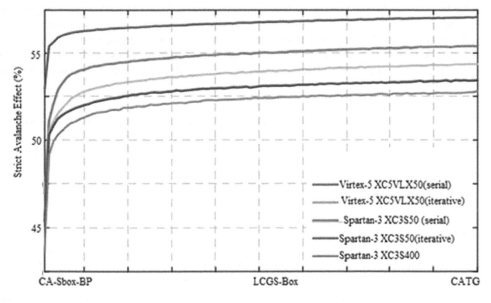

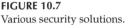

FIGURE 10.7
Various security solutions.

failure or disconnection. As a result, using the switch-controller link to attack the network may be a realistic strategy. Because the control and data planes are separated, an attacker can use OpenFlow rules to manipulate flows surreptitiously, resulting in active network attacks (Zhang, Beheshti, & Tatipamula, 2011).

10.6 Security Solutions of SDN Planes

In terms of network authentication, outlier detection, attack detection, and attack remediation are all areas where we specialize. SDN may improve the security level of any traditional network; yet, these benefits have created new risks and vulnerabilities. Figure 10.7 presents various security measures and solutions for securing the three SDN planes and are discussed as follows (Shu et al., 2016):

10.6.1 Application Plane Security Solutions

NICE is an OpenFlow application correctness checker which combines model checking with concolic execution in a novel technique for traversing the state space of unmodified controller programs developed for the popular NOX platform (Canini, Venzano, & Pere̊, 2012).

A novel strategy was created to solve the issue of scalability, which blends model verification with symbolic execution of event handlers. In addition, to decrease the state space, a simpler OpenFlow switch model was designed, as well as effective ways for creating event inter-leavings that are likely to detect weaknesses in OpenFlow applications. NICE incorporates symbolic execution, model verification, and search approaches. When applied to basic instances, it is five times faster and has a small overhead.

Son, Shin, Yegneswaran, Porras, & Gu (2013) built FLOVER, an OpenFlow application that worked on a controller and verified that no new rules produced by the controller violated a set of established attributes. The authors produced a prototype of the flow verification tool (FLOVER), which turns a flow table into a sequence of Yices statements and then evaluates if these assertions violate a network security policy. FLOVER can be used in two ways:

i. In-line mode, in which FLOVER checks the flow rule after each modification in the flow rule, and
ii. Batch mode, where the controller is verified regularly to enhance response time.

It has been determined that the performance of this method may disclose rule adjustments and violations for up to 200 rules in less than 120 milliseconds and 131 milliseconds, respectively.

FRESCO is an OpenFlow security application framework built by Seugwon Shin et al. (2013), which comprises a library of modules that can be used to build a whole cybersecurity application or output flow rules for each danger recognized by a detection module. This solution employs an OpenFlow application to provide a Click-inspired programming platform that allows security researchers to design, distribute, and combine a variety of vulnerability detection and mitigation tools. FRESCO has been proved to have a low overhead, allowing for the quick construction of common security tasks with a lot less lines of code. This technology provides a solid new platform for developing and deploying novel security solutions in the present era.

VeriCon is a tool that is used to verify the correctness of controller software and detect problems in SDN applications on a broad scale across all topologies and sequences of network events, according to Ball et al. (2014). This technique validates the correctness of each single network event in relation to the defined invariant, allowing it to grow to accommodate enormous applications. It provides a distinct technique for inferring invariants, which has been proved to be effective on basic controller systems, to relieve the programmer of the effort of providing inductive invariants and is considered as a first step toward realistic tools for validating SDN applications for network-wide invariants. The promising potential of this approach for providing the functional correctness of controller programs of interest encourages more study in this area.

Wundsam, Levin, Seetharaman, & Feldmann (2019) devised a method known as **OFRewind** that can capture and replay specified network data in order to track for network irregularities. It is a tool that assists operators in reproducing software faults, identifying data-path limits, and locating network problems. It allows to control the topology and traffic that is captured and then replayed during a debugging process. This framework is small enough that it may be turned on by default in production networks. This can keep the flow order, and the timing is precise enough for a variety of applications. On the other hand, further developments would increase its applicability. Researchers may, however, collect and maintain a consistent collection of traces in the future, which will be utilized as input for automated regression tests as well as benchmarks for new network component testing. Furthermore, it is expected that this strategy would be helpful in resolving production challenges in ongoing OpenFlow deployment efforts.

Elazim, Sobh, & Bahaa-Eldin (2019) designed a fine-grained permission system known as **PermOF** that uses a thread-based isolation technique with a customizable permission list. OF controller is vulnerable to privilege abuse due to involvement of third party, which allows for a series of control-plane attacks that might compromise the entire network. To reduce this hazard, the different possible remedies were devised before focusing on reducing the privilege of OF applications. This authorization system provides access control tokens that can be used to allow or reject any application. It contains four permission types which includes:

i. Read permissions, which allow to view and handle any sensitive application data.
ii. Notification permissions, which allow the program to be notified of certain occurrences.
iii. Write permissions allow you to change the state of a controller or a switch.
iv. System permissions, which grant access to the operating system's local resources for the application.

10.6.2 Control Plane Security Solutions

OF-GUARD is a lightweight design for SDN networks presented by Wang, Xu, & Gu (2015), which uses two software modules to minimize data-to-control-plane saturation assaults: packet migration and data-plane caching. To govern the pace at which packets are delivered by the data layer, packet migration uses a rotating scheduling strategy depending on the network's utilization ratio. Table-miss packets are sent to the data-plane cache (DPC), which includes proactive flow rules, caches table-miss packets, and distinguishes false from genuine packets, enabling for the detection of a flooding attack risk. However, as the number of assaults grows, bandwidth becomes scarce. It protects the whole network, but it does so at the expense of delay.

Seunghyeon Lee, Kim, Shin, Porras, & Yegneswaran (2017) developed Athena, a software solution based on the SDN protocol with a well-structured development interface and general-purpose features for quickly generating a variety of anomaly detection services and network monitoring activities with low programming effort. It is a distributed system application hosting architecture that provides for unparalleled scalability when compared to existing SDN security monitoring and analysis attempts. It is designed to run on current SDN infrastructures, making it simple and cost-effective to implement. It accumulates and develops network features in a distributed way above the SDN controller instances and then broadcasts the network characteristics to a distributed database. It has a machine learning (ML) architecture that can be used to build anomaly detection techniques and then deploy them as tasks throughout the computer cluster to speed up the construction of real-time detection models. It provides high-level APIs that enable operators to quickly create and deploy anomaly detection apps with little code. This method cuts the total calculation time for anomaly detection while simultaneously improving the scalability of the data management systems that execute these algorithms.

Fernandez (2013) proposed a robust controller that combines reactive and proactive techniques, with proactive controllers recognized for their scaling capability and reactive controllers renowned for their comprehension of traffic flow. The composite controller is a reactive controller that configures routes on request, but it can also adjust proactively and predict traffic patterns, allowing it to serve as a proactive controller and enhance overall system performance. This solution solves the scalability problem by providing a new intelligent switch that creates paths on-demand, improving the performance of proactive methods while keeping reactive methods functioning. The controller serves as a learning switch that does not require management interaction, but it may also be utilized to build up a proactive Reinforcement Learning system. The results show that by minimizing the number of control messages, the performance of a controller has improved. Certain algo-rithms, on the other hand, will be able to automatically configure the paths, increasing speed while keeping the configuration simple. However, in order to improve system man-ageability, researchers should focus on the Reinforcement Learning approach for design-ing routes based on traffic behavior in the future.

Phemius, Bouet, & Leguay (n.d.) suggested DISCO in order to overcome single point of failure in multidomain SDN networks, an open distributed SDN control plane has been devel-oped. This architecture is made up of two modules: (i) intra-domain module, which combines the controller's major functions; it regulates and monitors the network, and it dynamically reacts to network problems to deflect any potentially dangerous traffic; (ii) inter-domain mod-ule, which manages communication with other DISCO controllers. The inter-domain module is made up of two modules: (i) Messenger and (ii) Agents. Messenger establishes a communi-cation route between the controllers, allowing them to share information and make requests. Message orientation, priority queuing, routing, consistent delivery, and security are all ele-ments of the advanced message queuing protocol (AMQP), which is utilized in messenger communication. Agents aid with reservation procedures and QoS routing.

According to Voellmy & Wang (2012), McNettle is an extensible SDN control system that supports control algorithms and requires globally observable state changes occurring at flow arrival rates, with control event processing speed scaling with the number of sys-tem CPU cores. This is a simple and scalable SDN controller design that uses OpenFlow switches to improve control server performance and decrease latency while maintaining global visibility. Even the most demanding network control applications can be handled by these multicore architectures, which have ample memory, input/output, and process-ing resources. It is used to track and monitor Internet flow as well as respond to any harm-ful efforts that may arise. It works better with a constant number of cores, but as the number of switches increases, throughput decreases. As the number of switches exceeds 500, latency reduces as the number of cores increases.

HyperFlow, an OpenFlow control plane with a distributed event-based control plane, was suggested by Tootoonchian and Ganjali (2010). HyperFlow is theoretically centralized yet physically dispersed, allowing for scalability while maintaining the advantages of cen-tralized network control. It also makes it possible to join separately managed OpenFlow networks, which is presently unavailable in most OpenFlow installations. OpenFlow switches serve as data forwarding and event propagation methods, whereas NOX control-lers act as decision components. A specific controller is allocated to each switch. When a controller fails, the switches that are attached to it must be modified to link to a neighbor-ing controller. It employs a messaging system that necessitates event storage, maintains the sequence of events broadcast by the same controller, and allows each partition to function independently in disseminating events from the controller to other partitions. It ensures that all controllers can see the whole network. It is anticipated that it reduces reaction time by half by halving the time it takes to produce and install new flow rules.

10.6.3 Data-Plane Security Solutions

FortNOX is a software innovation established by Porras et al. (2012) that allows OF secu-rity applications to set enforceable flow limits. A rule conflict detection engine is used, and requests for flow rule insertion are sent via it. When a conflict is detected, the rule insertion user with high security authority can accept or reject the new rule. This strategy is a vital first step toward making OF networks more secure, but more work has to be done in terms of developing complete features that span a wide range of security services.

Khurshid, Zou, Zhou, Caesar, & Brighten Godfrey (2013) devised VeriFlow as a solution for monitoring flow rules in real time and avoiding erroneous rules from generating connection problems. It is utilized to preserve network performance while reducing inspection delay. VeriFlow is a layer that stands between controllers and switches that checks for dynamic violations when rules are added, updated, or removed. It includes an API for testing custom invariants and can check multiple header fields. According to a prototype implementation connected to the NOX OpenFlow controller and powered by a Mininet OpenFlow network and Route Views trace data, it can undertake rigorous verification within hundreds of microseconds per rule insertion or deletion. VeriFlow is assumed to be a tool that does real-time network-wide invariant verification as well as real-time forwarding table updates.

DELTA is a revolutionary SDN security evaluation system suggested by Seungsoo Lee et al. (2017). It is capable of automatically executing attack scenarios against SDN modules and identifying security issues. It was an important first step toward developing a method for automatically evaluating critical data flow exchanges between SDN components in search of known and potentially unknown vulnerabilities. This framework, which was built for OpenFlow-enabled networks and has been upgraded to work with the most prevalent OpenFlow controllers on the market, is the first and only SDN-focused security assessment tool presently available. Also introduced was a DELTA-integrated generalizable SDN-specific black-box fuzz testing technique. This fuzz testing approach may be used by the operator to do in-depth testing of the data input processing logic of a number of OpenFlow component interfaces.

Kreutz, Yu, Esteves-Verissimo, Magalhaes, and Ramos (2017) proposed KISS, which stands for Keep It Simple Stupid. A secure communication architecture with unique key distribution techniques and secure channel support was a first step toward tackling security concerns while minimizing cost and complexity. It was a replacement for existing systems based on traditional TLS and PKI configurations, which aimed to improve the reliability of control communications and decreased the complexity of infrastructure. An integrated device verification value (iDVV), a predictable yet indistinguishable-from-random secret code generating approach, was the system's most novel component. In terms of both robustness and simplicity, iDVV is expected to make a significant contribution to SDN's authentication and secure communication concerns.

The "FlowChecker" configuration verification tool was designed by Al-Shaer and Al-Haj (2010), and it is used to find internal switch misconfiguration. By encoding FlowTables configuration using binary decision diagrams, it employs the model checker approach to simulate a linked network of OpenFlow switches. It is a centralized server application that takes OpenFlow application requests, analyzes and troubleshoots the setup, and reports back. It can work as a standalone service, communicating with OpenFlow controllers through a secure SSL channel. The controller sends FlowChecker queries with information on the topology, users, controller IDs, and name binding. According to the time performance research, administrators may detect FlowTable misconfigurations in real time, offering them a strong tool to address any conflicts that may emerge across numerous OpenFlow switches. The CTL language enables more complex query writing to validate certain properties or retrieve data for QoS analysis.

Sherwood et al. (2009) suggested FlowVisor, a novel switch virtualization solution in which the same hardware forwarding plane may be used by several logical networks, each with its own forwarding logic. A network virtualization layer sits between switches and controllers and controls the underlying physical network using the OpenFlow protocol. A research platform was built using switch-level virtualization to allow several network experiments to run concurrently with actual traffic while maintaining isolation and hardware forwarding rates. This method makes use of common switching chipsets and does not necessitate the use of programmable hardware. This method is also said to increase overhead in operations at the control and data forwarding levels. Furthermore, the FlowVisor adds 16 milliseconds to the delay for receiving flow messages from the switch to the controller, and this latency will be 16 milliseconds for sensitive applications, a significant overhead over the system; without the FlowVisor, the overhead will be 12 milliseconds.

10.7 Software-Defined Networking in Cyber-Physical Systems (CPS)

A CPS is made up of a number of sensors or sensor-based devices that collect data from a variety of sources. It is made up of self-contained units that communicate often. A few examples of CPS-based systems are smart homes, smart grids, smart cities, mobile ad hoc networks (MANETs), and vehicular ad hoc networks (VANETs). In an edge network, CPS frequently has execution nodes. They use edge node computing resources as surrogates for task execution on their resource-constrained devices. When CPS is used in this way, resource constraints in CPS execution are removed. As a result of the utilization of edge, CPS adoption and capabilities continue to rise. CPS confronts a number of design and implementation issues, including (Kathiravelu & Veiga, 2017):

- Instability of the execution environments,
- Interoperability within the system,
- In the case of a system or network failure, security, worldwide fault-tolerance, and restoration are all important considerations,
- Controlling large-scale geo-distributed execution environment is a challenge,
- System modeling and design, and
- Intelligent agent management and orchestration.

By isolating the control layer as a unified controller from the dispersed network's data forwarding parts, SDN provides reusability and administration capabilities, among many other benefits, to networks. Recent researchers have examined using SDN to help with CPS deployment. To increase the robustness of multinetworks in CPS, SDN has been proposed. Through SDN-assisted emulations, SDN has been used to protect CPS networks and increase their resilience (Ahmed, Blech, Gregory, & Schmidt, 2016).

10.7.1 Software-Defined Cyber-Physical Networks: Challenges and Benefits

The various benefits and drawbacks of SDN-enabled cyber-physical networks are discussed as follows (Molina & Jacob, 2018):

10.7.1.1 Manageability

The growing number of networked devices necessitates the use of advanced management solutions. In order to implement complicated high-level network regulations, traditional approaches typically need network operators to deal with low-level vendor-specific configurations. The management of software-defined networks, on the other hand, is greatly simplified because all parts are handled via standard protocols. As a result, SDN enables network programmability, which is becoming a need for future networks.

Rinaldi, Ferrari, Brandao, and Sulis (2015) used SDN technology to create heterogeneous smart-grid networks. Experiments and literature were used to analyze the potential of SDN technology as a solution to simplify and automate network engineering in smart grids (SG). A specific assessment of an SDN system for SG may be required using extensive experimental efforts and modeling.

10.7.1.2 Resource Allocation

Routing problems that are difficult to solve benefit from a distinct control logic that takes a global perspective of the network. As a result, a controller can determine the best resource allocation. Following the discovery phase, the SDN controller is in charge of traffic engineering in CPS networks, defining the routing path between senders and recipients for unicast and multicast streams.

Huth & Houyou (2013) described the system architecture for network virtualization in industrial networks. Enterprise and telecommunication concepts have been mapped and used to meet industrial needs. The Slice Manager, a domain controller, directly manipulates a potentially heterogeneous network while providing planning and administration applications with a simple abstract view. Applications run on a virtual network called a "slice," which has unambiguous QoS assurances and bandwidth policies.

10.7.1.3 Time-sensitive SDN

To enforce latency constraints, a time-aware network controller should gather critical performance measurements offered by network devices, such as forwarding latency. This information might be used to arrange data pathways, plan traffic, and reserve bandwidth for time-sensitive streams. However, according to the CPS Public Working Group (PWG), SDN technologies may have a harmful effect on timing performance and the impact of SDN on timing performance should be thoroughly investigated.

Huang, He, Duan, Yang, and Wang (2014) investigated admission control using flow aggregation for QoS provisioning in SDN. Here, a model was created for admission control with flow aggregation and established analysis tools for estimating the needed quantities of bandwidth and buffer space that must be allocated at switches to meet performance criteria for delay and packet loss. The modeling and analysis built may be used to a wide range of OpenFlow-based SDNs with different implementations. According to both analytical and numerical research, flow pooling appears to boost bandwidth and buffer utilization in OpenFlow switches, as well as the scalability of an OpenFlow controller.

10.7.1.4 Industrial-grade Reliability

According to SDN standards for enhancing mission-critical CPS dependability, a controller can provide failover solutions via populating the forwarding tables in advance, with pre-installed instructions serving as backup rules, or on-demand by network devices. An SDN controller may make suitable routing decisions based on mobile node position and movement data.

Huang, Yan, Yang, Pan, and Liu (2016) presented an expanded OVS-based SDN-based vehicle sensor network architecture. It was an effort to make the design flexible and execute it in an expanded OVS environment. This approach can accomplish seamless handoffs when the scenario state changes and keeps car sensor networks' throughput steady. Of course, this design ensures that a packet is correctly processed and delivered even if connectivity with the SDN controller is lost. In precision agriculture, the method can lessen the impact of controller failure and increase the stability and survivability of SDN-based vehicle sensor networks. More importantly, the principles of stateful matching and a packet-processing service launched in the data plane might be applied to more agricultural settings.

10.7.1.5 Network Security

In terms of security, an external controller can implement filtering rules that protect cyber-physical networks from harmful assaults, in addition to the previously mentioned traffic isolation. Dynamic security policies in an SDN architecture can limit communication routes to only those necessary by control applications. As the number of authorized connections reduces, the number of possible attacks in CPSs diminishes.

Antonioli and Tippenhauer (2015) presented CPS emulation tools and explored how an OpenFlow-based control may help identify and avoid assaults. MiniCPS was proposed which combines Mininet, a physical-layer API, and a collection of matching component simulation tools to provide a flexible and lightweight simulation solution for CPS networks. This offers a basic API for simulating physical-layer interactions and can be utilized to design assaults and countermeasures that are directly relevant to actual systems.

10.7.1.6 Interoperability and Standardization of SDN

Standard interfaces and protocols are obviously necessary to connect collaborative CPS environments, such as autonomous automobiles engaging with intelligent transportation systems. Open standards, in fact, decrease dependency on vendor-specific equipment and allow for smooth CPS integration. The NIST Framework and Roadmap for Smart-Grid Interoperability, for example, has identified important protocols and organized them by the primary area to which they apply.

10.8 Conclusions and Future Work

SDN will become the new benchmark for networks as the future of networking becomes increasingly reliant on software. On the other side, there is a serious security risk with the SDN controller and applications that must be addressed before the SDN can be implemented safely. SDN deployment in production networks is still a concept and more work in SDN security is still to be done before this vision can become a reality. There appears to be important security issues to solve, as well as latent ones that have yet to be discovered, so the study area is still wide open, and every new contribution contributes to closing the gap between that vision and reality.

In addition, the aim of this chapter is to provide information to readers with a didactic introduction to SDN. More precisely, the chapter leads the reader through the evolution of

SDN and offers a straightforward and succinct description of SDN and proposed architecture. It also looks at the OpenFlow for SDN, as well as addressing various security challenges in SDN. The chapter offers a comprehensive look at promising SDN security issues at various planes of SDN architecture and a brief overview of SDN in CPS. It discusses the flaws in an open interface as well as the techniques for detecting and resolving these security concerns.

The future research directions includes exploring the new security issues and solutions for SDN, while offering network security solutions. Also, it is suggested that the undiscovered areas of SDN in CPS are identified.

References

Ahmad, I., Namal, S., Ylianttila, M., & Gurtov, A. (2015). Security in software defined networks: A survey. *IEEE Communications Surveys and Tutorials, 17*(4), 2317–2346. https://doi.org/10.1109/COMST.2015.2474118

Ahmed, K., Blech, J. O., Gregory, M. A., & Schmidt, H. (2016). Software defined networking for communication and control of cyber-physical systems. *Proceedings of the International Conference on Parallel and Distributed Systems— ICPADS, 2016-Janua*, 803–808. https://doi.org/10.1109/ICPADS.2015.107

Al-Shaer, E., & Al-Haj, S. (2010). FlowChecker: Configuration analysis and verification of federated OpenFlow infrastructures. *Proceedings of the ACM Conference on Computer and Communications Security*, 37–44. https://doi.org/10.1145/1866898.1866905

Ali, S., Alvi, M. K., Faizullah, S., Khan, M. A., Alshanqiti, A., & Khan, I. (2020). Detecting DDoS attack on SDN due to vulnerabilities in OpenFlow. *2019 International Conference on Advances in the Emerging Computing Technologies, AECT 2019*. https://doi.org/10.1109/AECT47998.2020.9194211

Antonioli, D., & Tippenhauer, N. O. (2015). MiniCPS: A toolkit for security research on CPS networks. *CPS-SPC 2015 – Proceedings of the 1st ACM Workshop on Cyber-Physical Systems-Security and/or Privacy, Co-Located with CCS 2015*, 91–100. https://doi.org/10.1145/2808705.2808715

Ball, T., Bjørner, N., Gember, A., Itzhaky, S., Karbyshev, A., Sagiv, M., … Valadarsky, A. (2014). VeriCon: Towards verifying controller programs in software-defined networks. *Proceedings of the ACM SIGPLAN Conference on Programming Language Design and Implementation (PLDI)*, 282–293. https://doi.org/10.1145/2594291.2594317

Bannour, F., Souihi, S., & Mellouk, A. (2018). Distributed SDN control: Survey, taxonomy, and challenges. *IEEE Communications Surveys and Tutorials, 20*(1), 333–354. https://doi.org/10.1109/COMST.2017.2782482

Canini, M., Venzano, D., & Pereˇ, P. (2012). A NICE way to test OpenFlow application nsdi12-final105. pdf. *USENIX Symposium on Networked Systems Design and Implementation*.

Correa Chica, J. C., Imbachi, J. C., & Botero Vega, J. F. (2020). Security in SDN: A comprehensive survey. *Journal of Network and Computer Applications, 159*, 102595. https://doi.org/10.1016/J.JNCA.2020.102595

Dargahi, T., Caponi, A., Ambrosin, M., Bianchi, G., & Conti, M. (2017). A survey on the security of stateful SDN data planes. *IEEE Communications Surveys and Tutorials, 19*(3), 1701–1725. https://doi.org/10.1109/COMST.2017.2689819

Deshpande, H. A. (2013). Software defined networks: Challenges, opportunities and trends. *International Journal of Science and Research, 4*, 2319–7064. Retrieved from www.ijsr.net

Elazim, N. M. A., Sobh, M. A., & Bahaa-Eldin, A. M. (2019). Software defined networking: attacks and countermeasures. *Proceedings – 2018 13th International Conference on Computer Engineering and Systems, ICCES 2018*, 555–567. https://doi.org/10.1109/ICCES.2018.8639429

Feamster, N., Rexford, J., & Zegura, E. (2014). The road to SDN: An intellectual history of programmable networks. *Computer Communication Review, 44*(2), 87–98. https://doi. org/10.1145/2602204.2602219

Fernandez, M. P. (2013). Comparing OpenFlow controller paradigms scalability: Reactive and proactive. *Proceedings – International Conference on Advanced Information Networking and Applications, AINA,* 1009–1016. https://doi.org/10.1109/AINA.2013.113

Foschini, L., Mignardi, V., Montanari, R., & Scotece, D. (2021). An SDN-enabled architecture for IT/OT converged networks: A proposal and qualitative analysis under DDoS attacks. *Future Internet, 13,* 258. https://doi.org/10.3390/fi13100258

Fraser, B., Sezer, S., Scott-Hayward, S., Chouhan, P.K., Lake, D., Finnegan, J., Viljoen, N., Miller, M., & Rao, N. (2013). Are we ready for SDN? Implementation challenges for software-defined networks. *IEEE Communications Magazine, 51,* 36–43.

Hayajneh, A. Al, Bhuiyan, M. Z. A., & McAndrew, I. (2020). Improving internet of things (IoT) security with software-defined networking (SDN). *Computers, 9*(1). https://doi.org/10.3390/computers9010008

Hu, F., Hao, Q., & Bao, K. (2014). A survey on software-defined network and OpenFlow: From concept to implementation. *IEEE Communications Surveys and Tutorials, 16*(4), 2181–2206. https://doi.org/10.1109/COMST.2014.2326417

Huang, J., He, Y., Duan, Q., Yang, Q., & Wang, W. (2014). Admission control with flow aggregation for QoS provisioning in software-defined network. *2014 IEEE Global Communications Conference, GLOBECOM 2014,* 1182–1186. https://doi.org/10.1109/GLOCOM.2014.7036969

Huang, T., Yan, S., Yang, F., Pan, T., & Liu, J. (2016). Building sdn-based agricultural vehicular sensor networks based on extended open vSwitch. *Sensors (Switzerland), 16*(1), 1–17. https://doi. org/10.3390/s16010108

Huth, H.-P., & Houyou, A. M. (2013). Resource-aware virtualization for industrial networks. *Dcnet,* 44–50.

Kathiravelu, P., & Veiga, L. (2017). SD-CPS: Taming the challenges of Cyber-Physical Systems with a software-defined approach. *2017 4th International Conference on Software Defined Systems, SDS 2017,* 6–13. https://doi.org/10.1109/SDS.2017.7939133

Khurshid, A., Zou, X., Zhou, W., Caesar, M., & Brighten Godfrey, P. (2013). VeriFlow: Verifying network-wide invariants in real time. *Proceedings of the 10th USENIX Symposium on Networked Systems Design and Implementation, NSDI 2013,* 15–27.

Kreutz, D., Yu, J., Esteves-Verissimo, P., Magalhaes, C., & Ramos, F. M. V. (2017). The KISS principle in software-defined networking: An architecture for keeping it simple and secure. Retrieved from http://arxiv.org/abs/1702.04294

Lee, Seunghyeon, Kim, J., Shin, S., Porras, P., & Yegneswaran, V. (2017a). Athena: A framework for scalable anomaly detection in software-defined networks. *Proceedings – 47th Annual IEEE/IFIP International Conference on Dependable Systems and Networks, DSN 2017,* 249–260. https://doi. org/10.1109/DSN.2017.42

Lee, Seungsoo, Yoon, C., Lee, C., Shin, S., Yegneswaran, V., & Porras, P. (2017b). DELTA: A security assessment framework for software-defined networks, (March). https://doi.org/10.14722/ndss.2017.23457

Molina, E., & Jacob, E. (2018). Software-defined networking in cyber-physical systems: A survey. *Computers and Electrical Engineering, 66,* 407–419. https://doi.org/10.1016/j.compeleceng.2017.05.013

Mubarakali, A., & Alqahtani, A. S. (2019). A survey: Security threats and countermeasures in software defined networking. *2019 IEEE 2nd International Conference on Information and Computer Technologies, ICICT 2019,* (March), 180–185. https://doi.org/10.1109/INFOCT.2019.8711319

Nunes, B. A. A., Mendonca, M., Nguyen, X. N., Obraczka, K., & Turletti, T. (2014). A survey of software-defined networking: Past, present, and future of programmable networks. *IEEE Communications Surveys and Tutorials, 16*(3), 1617–1634. https://doi.org/10.1109/SURV.2014.012214.00180

Phemius, K., Bouet, M., & Leguay, J. (n.d.). DISCO : Distributed Multi-domain SDN Controllers.

Porras, P., Shin, S., Yegneswaran, V., Fong, M., Tyson, M., & Gu, G. (2012). A security enforcement kernel for OpenFlow networks. *HotSDN'12 – Proceedings of the 1st ACM International Workshop on Hot Topics in Software Defined Networks*, 121–126. https://doi.org/10.1145/2342441.2342466

Pradhan, A., & Mathew, R. (2020). Solutions to vulnerabilities and threats in software defined networking (SDN). *Procedia Computer Science, 171*(2019), 2581–2589. https://doi.org/10.1016/j.procs.2020.04.280

Prathima Mabel, J., Vani, K. A., & Rama Mohan Babu, K. N. (2019). SDN security: Challenges and solutions. *Lecture Notes in Electrical Engineering, 545*, 837–848. https://doi.org/10.1007/978-981-13-5802-9_73

Rana, D. S., Dhondiyal, S. A., & Chamoli, S. K. (2019). Software Defined Networking (SDN) challenges, issues and solution. *International Journal of Computer Sciences and Engineering, 7*(1), 884–889. https://doi.org/10.26438/ijcse/v7i1.884889

Rinaldi, S., Ferrari, P., Brandao, D., & Sulis, S. (2015). Software defined networking applied to the heterogeneous infrastructure of Smart Grid. *IEEE International Workshop on Factory Communication Systems – Proceedings, WFCS, 2015-July.* https://doi.org/10.1109/WFCS.2015.7160573

Scott-Hayward, S., O'Callaghan, G., & Sezer, S. (2013). SDN security: A survey. *SDN4FNS 2013–2013 Workshop on Software Defined Networks for Future Networks and Services.* https://doi.org/10.1109/SDN4FNS.2013.6702553

Sherwood, R., Gibb, G., Yap, K., Appenzeller, G., Casado, M., McKeown, N., … Appenzeller, G. (2009). FlowVisor: A Network Virtualization Layer.

Shin, Seugwon, Porras, P., Yegneswaran, V., Fong, M., Gu, G., Tyson, M., … Park, M. (2013a). FRESCO : Modular Composable Security Services for Software-Defined Networks, 2(February).

Shin, Seungwon, Xu, L., Hong, S., & Gu, G. (2016). Enhancing network security through software defined networking (SDN). *2016 25th International Conference on Computer Communications and Networks, ICCCN 2016.* https://doi.org/10.1109/ICCCN.2016.7568520

Shin, Seungwon, Yegneswaran, V., Porras, P., & Gu, G. (2013b). AVANT-GUARD: Scalable and vigilant switch flow management in software-defined networks. *Proceedings of the ACM Conference on Computer and Communications Security*, 413–424. https://doi.org/10.1145/2508859.2516684

Shu, Z., Wan, J., Li, D., Lin, J., Vasilakos, A. V., & Imran, M. (2016). Security in software-defined networking: Threats and countermeasures. *Mobile Networks and Applications, 21*(5), 764–776. https://doi.org/10.1007/S11036-016-0676-X/TABLES/1

Son, S., Shin, S., Yegneswaran, V., Porras, P., & Gu, G. (2013). Model checking invariant security properties in OpenFlow. *IEEE International Conference on Communications*, 1974–1979. https://doi.org/10.1109/ICC.2013.6654813

Spooner, C.;, & Zhu, J. (2016). A review of solutions for SDN-exclusive security issues. *International Journal of Advanced Computer Science and Applications (IJACSA), 7*(8). https://doi.org/10.14569/IJACSA.2016.070817

Stancu, A. L., Halunga, S., Suciu, G., & Vulpe, A. (2015). An overview study of software defined networking. *Proceedings of the IE 2015 International Conference, 3*(April), 3053–3055. Retrieved from www.conferenceie.ase.ro

Tootoonchian, A., & Ganjali, Y. (2010). HyperFlow: A distributed control plane for OpenFlow. *2010 Internet Network Management Workshop/Workshop on Research on Enterprise Networking, INM/WREN 2010.*

Upc, F. I. B. (2012). OpenFlow : Enabling Innovation in Campus Networks Protocol.

Vaughan-Nichols, S. J. (2011). TECHNOLOGY NEWS OpenFlow: The Next Generation of the Network?

Voellmy, A., & Wang, J. (2012). Scalable software defined network controllers. *Computer Communication Review, 42*(4), 289–290. https://doi.org/10.1145/2377677.2377735

Wang, H., Xu, L., & Gu, G. (2015). FloodGuard: A DoS attack prevention extension in software-defined networks. *Proceedings of the International Conference on Dependable Systems and Networks, 2015-Septe*, 239–250. https://doi.org/10.1109/DSN.2015.27

Wen, X., Chen, Y., Hu, C., Shi, C., & Wang, Y. (2013). Towards a secure controller platform for OpenFlow applications. *HotSDN 2013 – Proceedings of the 2013 ACM SIGCOMM Workshop on Hot Topics in Software Defined Networking*, *13*, 171–172. https://doi.org/10.1145/2491185.2491212

Wundsam, A., Levin, D., Seetharaman, S., & Feldmann, A. (2019). Ofrewind: Enabling record and replay troubleshooting for networks. *Proceedings of the 2011 USENIX Annual Technical Conference, USENIX ATC 2011*, 327–340.

Xia, W., Wen, Y., Foh, C. H., Niyato, D., & Xie, H. (2015, January 1). A survey on software-defined networking. *IEEE Communications Surveys and Tutorials*. Institute of Electrical and Electronics Engineers Inc. https://doi.org/10.1109/COMST.2014.2330903

Yao, G., Bi, J., & Guo, L. (2013). On the cascading failures of multi-controllers in software defined networks. *Proceedings – International Conference on Network Protocols, ICNP*, *1*, 3–4. https://doi.org/10.1109/ICNP.2013.6733624

Yoon, C., Lee, S., Kang, H., Park, T., Shin, S., Yegneswaran, V., … Gu, G. (2017). Flow wars: Systemizing the attack surface and defenses in software-defined networks. *IEEE/ACM Transactions on Networking*, *25*(6), 3514–3530. https://doi.org/10.1109/TNET.2017.2748159

Zhang, Y., Beheshti, N., & Tatipamula, M. (2011). On resilience of split-architecture networks. *GLOBECOM – IEEE Global Telecommunications Conference*. https://doi.org/10.1109/GLOCOM.2011.6134496

11

A Review on AI-ML Based Cyber-Physical Systems Security for Industry 4.0

Sima Das and Ajay Kumar Balmiki
Maulana Abul Kalam Azad University of Technology, Kolkata, West Bengal, India

Kaushik Mazumdar
IIT Dhanbad (ISM), India

CONTENTS

11.1 Introduction

Cyber-physical systems (CPS) [1–7] may be found as the mixture of data processing, verbal exchange, and somatic approaches, wherein embedded data processing and networking can manage physical systems and devices. The aggregate of devices, sensors, lodged algorithmic calculation intelligence, and conversation implementation permits CPS to display and control physical components through computer-based total algorithms strongly supported by way of net connectivity (IoT). Latest sensors with algorithmic skills authorize compatible and analytic implementation are fetching the center of CPS and protection-crucial systems. CPS additionally play an essential function in trade digitalization. Analogous strategies need computerized statistics accession and facts garage, beyond analyses on the cloud. Having ease of communication among physical devices and the cloud is a crucial

DOI: 10.1201/9781003241348-11

difficulty. Repeated upgrades in computational techniques, artificial intelligence (AI) [8], and communications, as well as sensor and actuators degrees, permit systems to filter their overall production interacting with the circumambient territory. Evolution and self-mastering need to be present on every occasion fault-tolerant operation and resiliency needs to be taken into consideration. Cyber security in operation is required whenever the management of a critical framework is considered, due to the high level of security required. Summarizing the above, it can be noticed that the present-day design of CPS has to stand several challenges whilst integrating the new AI methodologies, preserving security and resiliency at high tiers. Subjects of engrossment incorporate, but are not restricted to: Design methodologies of CPS and IoT; Formal evaluation and verification of CPS and IoT, safety-critical systems; Cybersecurity components, which include cryptographic algorithms, protocols, e-services, and many others.; The impact of AI-powered CPS and IoT security solutions on overall CPS performance; modeling of attack prediction; industrial digitalization; and CPS. CPS have appeared as a merging name for systems wherein cyber elements (i.e., the computing and communique components) and physical components are firmly incorporated, each in outline and at some stage in performance. Such systems use algorithmic calculation and transmission to enormously submerge in and interact with personal physical techniques in addition to augmenting present and adding new talents. As such, CPS is a combination of algorithmic, networking, and physical processes. Embedded computer systems and networks reveal and control the physical tactics, with remark loops wherein physical strategies have an effect on computations and vice versa. The economic and societal ability of such structures is vastly more than what has been realized, and fundamental investments are being made globally to increase the generation. AI paradigms for smart CPS makes a specialty of the latest advances in AI-based strategies toward affecting relaxed CPS. This e-book provides investigations on today's research issues, programs, and achievements within the discipline of computational intelligence paradigms for CPS. Completing the points that encompass independent structures, getting admission to control, system learning, and intrusion detection and prevention systems, this book is ideally designed for engineers, enterprise professionals, practitioners, scientists, managers, students, academicians, and researchers looking for modern studies on synthetic intelligence and CPS.

The rest of the chapter is as follows: Motivation and contribution is discussed in Section 11.2, features of AI-ML based CPS security for Industry 4.0 are discussed in Section 11.3, literature survey is discussed in Section 11.4, application, advantages, and disadvantages of machine learning for Industry 4.0 security are discussed in Section 11.5, and the last section discusses the conclusion and future work of the proposed system.

11.2 Motivation and Contribution

This chapter fabricates the work which gives the concept and idea regarding the need and availability of Industry 4.0; hence, it also conceptualizes the aim regarding Industry 4.0. But there are some facts and points that need to be managed in various industries and sectors, that is, security and protection. We need to improvise the security factor and hence also upgrade the system performance and fabrication. Various types of devices, systems, functions, and technologies are embedded, so management of threats and security becomes tougher, and thus it gives us the agenda and motivation to solve the issue of security and also the proper communication and engagement of devices in real time as well. If the engagement is good, the output will be accurate and also the system will become more secure and reliable.

11.3 Features of AI-ML Based Cyber-Physical Systems Security for Industry 4.0

In this section, we discuss features of AI-ML based CPS Security for Industry 4.0.

11.3.1 Industry 4.0

In the past couple of years, a fourth contemporary revolution has arisen, known as Industry 4.0. Industry 4.0 takes the emphasis on modern revolution from decades to an unprecedented off-track with the support of linking through the internet of things (IoT), acceptance of constant information, and the dispensation of modern physical frameworks. Industry 4.0 (Figure 11.1.) provides a better denial coverage, interconnectivity, and comprehensive method to handle or manage the processing. It is connected physically with modern technology and takes into consideration good collaboration and getting into divisions with collaborators, merchants, items, and others. Industry 4.0 captivates businessmen to more readily monitor and control every section of their work and allows them to utilize short-time details to help proficiency, development phases, and drive evolution.

Enterprise resource planning (ERP): Business managers manage tools that can be utilized to control demography all over a company. IoT stands for internet of things, an idea that gives us the explanation of linking among physical gadgets such as sensors and the Internet. IIoT stands for the industrial internet of things, an idea that belongs to the linking or interlinking among humans, records, and machines as they relate to manufacturing. Big information: large information belongs to a huge number of established or amorphous

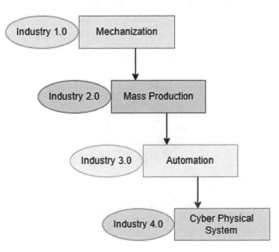

FIGURE 11.1
Evaluation of industry.

information which can be assembled, stored, organized, and analyzed to reveal styles, trends, associations, and opportunities. AI: AI is an idea that belongs to a PC's potential to execute duties and take judgment that might traditionally need several degrees of mankind's intellect. M2M stands for mobile-to-mobile and belongs to the communique that takes place among two distinct systems through Wi-Fi or stressed-out networks. Evolution and modernization belong to the manner of gathering and changing one kind of data or details into an advanced layout. Advance manufacturing unit: An advanced and digital factory spends in grip Industry 4.0 technology, resolution, and techniques. Cloud computing: Cloud computing belongs to the usage of remote servers provided on the Internet to save, operate, and maintain certainty. Instantaneous or real-time records operating: Instantaneous statistics operating belongs to the capabilities of a laptop blueprint, framework, and apparatus to continuously and routinely procedure testimony and fabricate instantaneous results and perception.

11.3.2 Cyber-Physical System

CPS (Figure 11.2.), also called as cyber production, belongs to an Industry 4.0 authorized construct habitat that fabricates instantaneous facts series, detecting, and lucidity over each factor of production execution. CPSs are integrations of computation, networking, and bodily procedures; they are the combination of various creations whose primary goal is to manipulate a somatic procedure and, with remarks, manage themselves to the most recent situations in real time. CPSs are remodeling, in the method just like mankind and systematic structure are linked together, just because the net has converted, the method by which mankind engages with statistics. People will always be essential in this situation. While the maximum bendy and smart "entity" is inside the CPS, humans anticipate the function is a type of "maximum-stage management instance", directing the execution of the commonly self-functioning and self-organizing approaches. A CPS, being made up of many distinct factors, calls for complicated fashions to summarize every sub-machine and its conduct. Vigorous inter-linked sub-structures are then orchestrated through an overarching model: a managed entity that ensures deterministic conduct of each sub-system. Current design gear wants to be upgraded to do not forget the interactions between the diverse sub-structures, their affiliations, and extraction. Overall, interaction leads to a state

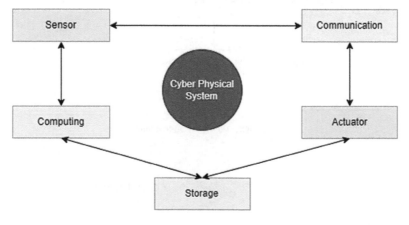

FIGURE 11.2
Attack surface of cyber-physical system.

of abeyance, range, and dependability, which is affected in large part by random communication between substructures. If we state non-wired communities, points such as machine area, producing conditions, and visitor burden exchange over time. This manner that the verbal exchange community also wishes to be summed up as one of the fashions in the general CPS "version of fashion". The time for which to perform a managed assignment may be essential to enable an accurate functioning device. Physical approaches are the mixture of several objects fabricated side by side. A model of time, which is static with the actuality of time dimension and time synchronization, needs to be systematized across all fashions. In the future, CPSs may be active in all business sectors and within the industry 4.0 paradigm. CPSs will fabricate production methodologies transforming the quality of tomorrow for the industry. Manufacturing territory could be self-configuration, self-management, and self-optimization, mainly to extra ability, comfortability, and fee reliance. As explained below, each practical factor of a manufacturing system might be pretentious, from architecture to fabricating, through the delivery system, and increasing the customer business and help. In this case, the destiny manufacturing facility might be a CPS, or a fixed of communicating CPSs, wherein particularly talented employees could have the perception of the execution directly from the coordinated wise system and an imperative managed entity. This manufacturing facility may be hyperlinked and statistics extensive, depending on a 100% comfy industry-grade 5G network. The cadence and technique at which agencies undertake modern evolution will vary. However, the line-by-line step distribution direction toward the full Industry 4.0 paradigm is particularly well explained. It includes six degrees, every block at the preceding one, as summarized below. In this direction, the primary two steps are automated and interconnected. A whole sum of automation, computation, networking, and somatic methods has not yet been reached at this point. In a third step, with the usage of systems and advanced modern fashions, the lucidity of technological facts and tactics is carried out in actual time. Modern and advanced twin fashions are part of the sport right here. With virtual fashions, business sectors can not only see what is happening, but also hit upon and comprehend it, achieving the fourth step of lucidity. Here it's important to purify and elucidate records by using complete fact detecting, in any respect factory levels. The prognosticating capability and anticipating are going to be the next steps, which is the fifth step. It is set in fabricated unique situations, anticipating the chance of prevalence, and being geared up for the possible consequences. With virtual fashions, business sectors are not the handiest to see what occurs, but additionally hit upon and understand, attaining the fourth step of lucidity. Here, it's important to purify and elucidate records by using complete facts detecting, in any respect factory levels. The prognosticating capability and anticipating are going to be the next step which is the fifth step. It is set fabricating unique situations, anticipating the chance of prevalence, and being geared up for the possible consequences. The digital models are then dynamically up to date. Industry 4.0, however, can be applied with the 6th and last step: adaptability. It is ready automatically to take adapting steps right away and, when needed, with no people involvement. To sum up everything we've talked about so far, research and development in areas like radio networks, the cloud, and systems for improving devices can help us understand the full potential of CPSs.

11.3.3 Machine Learning and Artificial Intelligence

Machine learning [9–15] is the idea that computers should use AI to figure out and improve how they talk to each other without being told or programmed to do so. A decade ago, the term "Industry 4.0" was created to describe the process of digitalization in the business world. Since then, there has been a clear increase in the number of groups focused on the deployment or execution of modern and digital technologies like IoT, blockchain, and all the branches of AI: machine learning, deep learning, cognitive intelligence, and so on.

Infinium Global Research details in a file the benefits of computation in Industry 4.0. The implementation of technology, which includes device getting to know the sector, donates to enhancing fabrication, production performance, and permits faster extra pliable, and extra well-organized operation. The record additionally indicates that the growth in people donating to the modernization of the sector or area is nourishing the business sector marketplace. In this path, the European Union is shifting ahead with a company step. In February 2020, the European Commission provided the "White Paper on AI". A joint strategy among all EU countries, summarized by its president, Ursula von der Leyen, aims to attract more than 20 billion euros in investment in AI over the next ten years. A general statistic is predicted to be outreach with the donation of the personnel quarter and the subsidizing of the states. Public funding will raise Industry 4.0 and modern advancement in the electronics sector, the adaptation of cloud computing technology, and the deployment of the modern manufacturing facility, according to analysts at Global Research. Organizations from exceptional sectors could also be able to benefit from the blessings of the software of technologies including machine learning inside the enterprise, but exceptionally, they will be the chunk of four strategic regions for this technology: ceramics, automobile, assembling, and power control and food. The business sectors to gain maximum from device learning. Companies inside the ceramics, automotive, energy control, and food and beverage markets are already cashing in on the benefits of imposing AI via machine-getting to know algorithms. They are enforcing a generation that permits them to anticipate negative and misguided behavior, optimize manufacturing techniques, and analyze the marketplace or request in intensity which will realize it better and therefore adapt greater exactly to consumer needs. Following are the application areas of AI and IoT that enable Industry 4.0.

- Ceramics: In the ceramics [30] field, AI is starting to take part in a prime function. Profits of the system getting to know in the ceramics region. Machine learning algorithms are always being utilized, mainly in excellent manipulation programs. With numerous algorithms, it is feasible to anticipate the behavior of the fabric below high temperature and to identify the oddity and insufficiency in the tiles. The procedure is performed with the assistance of AI to find the abnormal or typical behaviour of substances all throughout the production method, turn it viable to manage and utilize the additives that come up with better resistance situations than those currently being synthetic. By spotting wrong styles, they may be capable of discovering oddities in goods early, decreasing abatement, and growing beneficially. Now, we almost discovered agencies that can be working with this era and are using it in this line or others. They are, chiefly, groups inside the ceramic, porcelain stoneware, and floors regions.
- Automotive: In the car zone, AI is likewise an era that is highly being used to develop business methods. The car and all correlated region are using machine learning to boom their yield. The profits of using machine learning are starting to be known inside the automobile sector. This enterprise uses such technologies to perform the predictive evaluation of element sturdiness and inside the early identification of oddities and issues. Another utility of ML in the car zone is the most effective of the delivery system. ML technologies constitute a golden possibility to develop

the production procedure of organizations within the car area. In this experience, they manage well, between different capabilities, the stock ranges wished inside the exclusive provisions. Most companies within the concerned vehicle zone are taking profit advantage of system studying to develop their fabrication strategies.

- Installations and power control: In the installations and strength control quarter, AI, via device studying, is selling tremendous advancement. Profits of system studying in the assembling and electricity efficiency quarter. The creation of these generations in this region is evolving smart networks or smart grids. As per the Business Insider portal, this form of the network would take the benefits of gadget mastering generation to discover actual-time calculation to better regulate the strength delivered to request with the aid of identifying intake design and to cut off any disasters or interruption that might occur at some point of the supply system. Another advancement in power computing might be associated with enhancing the working and boosting of the community, the drop-in provider, fee development, improvement anticipation via areas, the detection of consumption, and call for peaks or the behavior of positive customers or cities. The deployment of AI technology inside the electricity management of towns comes with special benefits to both people and agencies. According to take a survey with the aid of Juniper Research, smart grids will save residents a few $14 billion in strength fees by 2022. Lots of companies inside the zone are already harvesting these blessings, developing the power control of cities by using advanced gadgets and gaining knowledge of structures.

- In the food industry [31], AI through system gaining knowledge of algorithms is donating to a discount in charges and a development in high-quality. It is doing the same in all sectors, in the food and beverage enterprise and inside the catering enterprise. Profits of devices studying inside the meals zone. Machine getting to know technology allows the industry to advantage of some of the key blessings to enhance its commercial enterprise. One of those blessings is an evaluation of the grocery store for you to have a know-how of patron developments and therefore manage what the client simply wants.

- There is also another chunk of the machine learning program connected to the development of purity in a production plant. It may be utilized to find out whether a system is unclean and needs to be wiped clean or to screen and detect the purity of all employees worried inside the manufacturing chain. Machine learning is also utilized in sectors or areas for making the most effective food and beverage delivery chain. Nowadays, numerous establishments in the meals zone might be benefiting from AI and extra mainly from devices getting to know (Figure 11.3).

11.3.4 Deep Learning

Deep learning [16–20] is a branch of system mastering that is completely built entirely on Artificial Neural Networks, as neural networks are going to mimic the human mind so deep mastering is also a form of mimicry of the human brain. An in-depth one is getting to understand that we do not want to specifically apply everything. The idea of Deep learning is not new. It has been around for multiple years now. It's the hype nowadays due to the fact earlier we did not have that much processing electricity and a whole lot of facts. As in the last 20 years, the processing power will increase exponentially, deep mastering and gadget mastering got here in the photograph.

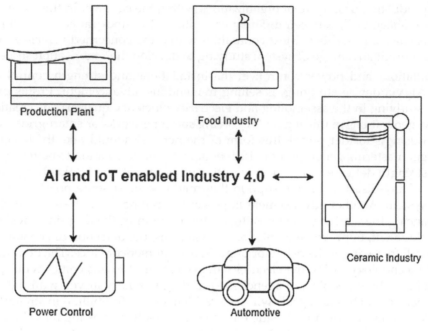

FIGURE 11.3
Application area of artificial intelligence.

It is to be explained that virtual evolution and application of computing strategy has been taking place in the area of the procurement industry for some time. As counterproductive troubled worldwide production inside the '60s and '70s, just about the entire large business enterprise was efficient and managed correct application like Toyota's Manufacturing Strategy. This sort of method trusted non-stop calculation and algorithmic computing of assembling of system variables and product capabilities. As the capacity and saving space of such data or details turn out to be advanced, computer systems were introduced for constructing those predictive fashions. This became the forerunner to fashionable advance anticipation of today. However, because the facts explosion maintains, conventional algorithmic computing does not hold up with such high-dimensional, non-established records broadcast. It is right here that deep studying glorifies and dazzles as it's naturally able to handle surprisingly nonlinear statistics styles and also permits you to find or recognize functions that are extraordinarily hard to be noticed using statisticians or statistics modelers non automatically. Quality Control in Machine Learning and Deep Learning. Machine getting to know, in preferred, and deep learning, specifically, can notably enhance the best manipulate obligations in a huge meeting line. In truth, analytics and ML-driven systems and nice development are anticipated to develop with the aid of 35%, and process visualization and automation are slated to grow via 34%, in line with Forbes. Conventionally, systems have only been powerful at recognizing first-class troubles with high-level metrics which include the weight or length of a product. Without going through a chance on extremely enlightened pc vision systems, it turned into not feasible to come across diffused optical hints on first-class problems at the same time as the components whizz with the aid of on a meeting line at a high pace.

11.4 Literature Survey

In this section, a literature survey has been discussed.

Further, the consistent, networked, and intelligent systems of industry 4.0 have increased the structures of cyber threats. In the cyber-physical systems (CPS) architecture and security requirements in Industry 4.0, a CPS testbed built on cloud computing and software-described network (SDN), or CPSTCS (Cyber-Physical System Testbed based on Cloud System), is proposed. The CPSTCS makes use of a network testbed based on cloud computing and SDN to recreate the cyber factors of CPS and real-time gadgets for the physical additives. The CPSTCS allows verifying cyberthreats in opposition to the virtual and physical dimensions of important infrastructures [21].

Industry 4.0 is a new idea; hence, threat assessment is essential. Numerous threat calculation strategies for Industrial Control Systems (ICS) and Industry 4.0 were anticipated, but it's tough to recognize influences on the physical global resulting from cyber-attacks in contradiction of ICS due to the fact that several are based on software simulations. They pay attention to the danger assessment and the use of ICS testbed that can benefit from resolving the overhead complications. In Industry 4.0, self-sustaining judgement and execution are required for the cyber-physical device; it's far primarily based on statistics and the use of artificial intelligence (AI) and cloud technology. Current studies examine cyber risks over assaults contrary to ICS with AI and cloud the usage of the ICS testbed. The proposed technique can clarify cyber dangers and effects on the actual international, and provide consistent countermeasures [22].

They [23] are presently going through the 4th manufacturing revolution in terms of cyber-physical systems. These structures are business computerization that allows several progressive functionalities via their networking and they get admission to the cyber international, hence converting our normal lives considerably. In this framework, innovative commercial enterprise fashions, painting methods, and improvement strategies that are currently not possible will get up. These changes may even strongly impact society and those who live in household life, shopping, and so forth. However, Industry 4.0 concurrently indicates features that constitute the challenges concerning the improvement of cyber-physical systems, dependability, safety, and record safety. After a brief introduction to Industry 4.0, this chapter describes a typical application that shows the most important parts.

Industry 4.0 denotes the 4th segment of the enterprise and industrial revolution, specific in that it runs on Internet-linked smart systems, with automatic factories, businesses, an improvement on request, and `just-in-time' improvement. Industry 4.0 consists of the combination of CPSs, IoT, cloud, and fog computation models for growing smart gadgets, smart houses, and smart cities. Given that Industry 4.0 comprises sensor fields, actuators, fog, and cloud processing paradigms, and community structures, manipulative appearances for essential challenges include: coping with varied foundations at the same time and maintaining security for a big, information-pushed system that is connected with the physical atmosphere. In this chapter, demanding situations are handled with the aid of a brand-new threat intelligence scheme that models the dynamic interactions of enterprise 4.0 additives, inclusive of physical and network structures. The organization contains two additives: a smart administration module and an intelligence module. The smart information control part connects to an Industry 4.0 machine to handle different record resources [24].

There [25] has been an excessive issue for security substitutes due to the current growth of cyber assaults, in particular, focused on important systems along with industry, scientific, and energy ecologies. Although the cutting-edge manufacturing organizations based on AI-pushed upkeep, predictions generated totally on corrupted information will absolutely result in damage to life and capital. An insufficient record safety mechanism can easily undertake the security and dependability of the community. The tested context obviates the lengthy-hooked-up certificate authority after enhancing the association of blockchain that decreases the fact processing put off and will increase price-powerful throughput.

Industry 4.0 has the ideas of the IoT in the manufacturing industry. Smart unions equipped objects that may be capable of communicating interconvertible and argument auto enhanced with other items. These manufacturing units are known as CPS. Behind the Industry 4.0-related examples of business robotic packages and CAE/CAD software by recognition on EPLAN P8, which predicts price with knowledge [26].

11.5 Advantages, Limitation Application Area

By embracing the many use cases for AI and device mastering in manufacturing, technology plant life can improve manufacturing excellence, predict fluctuations in market demand, reduce the number of significant incidents, increase their recognition for safety and environmental impact, and increase overall performance and productivity. Using ML and AI in production is a long-term process that will keep saving money and making more money for a long time.

11.5.1 Advantages of AI-ML Based Industry 4.0

Advantages of artificial intelligence (AI)- machine learning (ML) based Industry 4.0 are discussed below:

- Abatement of cost with the help of predictive maintenance results in much lower upkeep interest, identifying or fabricating less labour cost. It also reduces the wastage of listings and substances or stuff.
- Anticipating Remaining Useful Life (RUL): keeping tabs on the functioning of the system also results in growing situations that enhance performance whilst managing system fitness. Anticipating RUL minimizes the eventualities that cause unexpected interludes.
- Developed supply chain management via efficient stock control and an excellently observed and contemporized manufacturing waft.
- Independent machines and motors: They use self-sufficient winches and vehicles to speed up the process of getting boxes from trucks, ships, and other vehicles.
- Best Standard Management with active awareness to repeatedly improve the quality standard.
- Cooperation between developed and digital man-made devices improves worker safety and also boosts normal performance.
- Purchaser-centered production: It is very easy to change or adapt to the needs of the market.

11.5.2 Limitation of AI-ML Based Industry 4.0

The limitations of artificial intelligence (AI)-machine learning (ML) based Industry 4.0 are discussed below:

- Implementation instances, which may be prolonged relying on what you are attempting to enforce.
- Integration challenges and a lack of expertise in the present-day structures.
- Usability and interoperability with other structures and systems.

11.5.3 Application of AI-ML Based Industry 4.0

Application of AI and machine learning (ML) based Industry 4.0 are discussed below:

- Smart factories: Distinct through using their capacity to connect huge quantities of records, smart industries that are imposing Industry 4.0 are the use of AI in their manufacturing procedures. Constructors who have correctly endorsed a digital revolution may each establish and make use of their statistics to make use of AI and system mastering to progress satisfactory controller, adjustment, and preservation through analyzes of gadget capability and punctiliously reorganizing manufacturing facility traces. Though the blessings of AI are massive in terms of manufacturing strategies, it's critical to keep in mind that flowers must have an AI improvement plan in place, this is the concept of automation strategy to apply.
- Prediction: Statistics are accumulated in actual time to observe the system in predictive preservation circumstances. The purpose is to create a trace that could help predict and in the long run prevent screw-ups; increasingly more, through gaining knowledge of algorithms, AI organizations are being used to reap this aim. After analytical upkeep is computerized, plans can be more strategic when predicting the situation of the system and predicting while preservation must be achieved. Certainly, the operation of ML-based answers can cause a higher probability and the extended accessibility of the device.
- Computer vision (CV): From varieties of AI, CV is an area of data science that concentrates on allowing computer systems to become aware of and technique items in photos and motion pictures the identical way that human beings do. One of the riding elements in the back of the sphere's boom is the number of facts produced that are then used to teach and make laptop vision higher. Appreciation to developments in AI and modernizations in deep getting to know and synthetic neural networks, the sector has been intelligent to take remarkably in current centuries and has been unable to human beings in a few duties connected to predicting and classification objects.
- CPS: CPS entrench software into the physical international and seem in an inclusive program consisting of smart grids, robotics, and smart production. Quick developments in Internet-primarily based systems and applications have unlocked the opportunity for industries to operate the internet workspace to conduct effective and powerful day-to-day association from any place universal to offer a completely dispensed industrial atmosphere.
- Industrial robots: A business robot is a robotically managed, reprogrammable, versatile manipulator programmed. Distinctive applications encompass welding,

painting, ironing, meeting, pick and vicinity, and testing, which might be all done with excessive patience, pace, and exactness.

11.6 Conclusion and Future Scope

This chapter concludes the study of AI and machine learning based on security-related fields and sectors. This chapter also gives the concept of improvising Industry 4.0 in terms of manufacturing and also gives the idea of enhancing collaboration among diverse departments, giving the right statistics to be handed to the proper human beings on a real-time basis. This concept is not only in this sector but also applicable in the other sectors of security and protection. For fetching data in real-time, not only data acquisition but also proper measurement and action are required. Therefore, if this sector becomes weak, then this will impact the area of security and data protection as well. The aim is to facilitate and fabricate appropriate decision-making at the right time, thereby growing performance and productivity.

In the future, we can provide and implement this system to various industries and sectors just like food industries. The food industry also needs security and proper meaningful real-time data acquisition. The engagement of food industries with all the industries will also become better, and hence it, will improve the growth of this industry in terms of business, engagement, and security.

References

1. J. Jiang, "An improved cyber-physical systems architecture for Industry 4.0 smart factories," *2017 International Conference on Applied System Innovation (ICASI)*, 2017, pp. 918–920. DOI: 10.1109/ICASI.2017.7988589
2. Z. You and L. Feng, "Integration of Industry 4.0 related technologies in construction industry: A Framework of cyber-physical system," *IEEE Access*, vol. 8, pp. 122908–122922, 2020. DOI: 10.1109/ACCESS.2020.3007206.
3. J. Ruan, W. Yu, Y. Yang, and J. Hu, "Design and realize of tire production process monitoring system based on cyber-physical systems," *2015 International Conference on Computer Science and Mechanical Automation (CSMA)*, 2015, pp. 175–179. DOI: 10.1109/CSMA.2015.42
4. M. Ramadan, "Industry 4.0: Development of smart sunroof ambient light manufacturing system for automotive industry," *2019 Advances in Science and Engineering Technology International Conferences (ASET)*, 2019, pp. 1–5. DOI: 10.1109/ICASET.2019.8714236
5. W. Matsuda, M. Fujimoto, T. Aoyama, and T. Mitsunaga, "Cyber security risk assessment on Industry 4.0 using ICS testbed with AI and Cloud," *2019 IEEE Conference on Application, Information and Network Security (AINS)*, 2019, pp. 54–59. DOI: 10.1109/AINS47559.2019.8968698
6. L. Cavanini et al., "A preliminary study of a cyber-physical system for Industry 4.0: Modelling and co-simulation of an AGV for smart factories," *2018 Workshop on Metrology for Industry 4.0 and IoT*, 2018, pp. 169–174. DOI: 10.1109/METROI4.2018.8428334
7. M. Yu, M. Zhu, G. Chen, J. Li, and Z. Zhou, "A cyber-physical architecture for Industry 4.0-based power equipments detection system," *2016 International Conference on Condition Monitoring and Diagnosis (CMD)*, 2016, pp. 782–785. DOI: 10.1109/CMD.2016.7757942

8. G. Cheng, L. Liu, X. Qiang, and Y. Liu, "Industry 4.0 development and application of intelligent manufacturing," *2016 International Conference on Information System and Artificial Intelligence (ISAI)*, 2016, pp. 407–410. DOI: 10.1109/ISAI.2016.0092

9. W. Caesarendra, T. Wijaya, B. K. Pappachan, and T. Tjahjowidodo, "Adaptation to Industry 4.0 using machine learning and cloud computing to improve the conventional method of deburring in aerospace manufacturing industry," *2019 12th International Conference on Information & Communication Technology and System (ICTS)*, 2019, pp. 120–124. DOI: 10.1109/ICTS.2019.8850990.

10. H. Ouanan and E. H. Abdelwahed, "Image processing and machine learning applications in mining industry: Mine 4.0," *2019 International Conference on Intelligent Systems and Advanced Computing Sciences (ISACS)*, 2019, pp. 1–5. DOI: 10.1109/ISACS48493.2019.9068884

11. C. Weber and P. Reimann, "MMP – A platform to manage machine learning models in Industry 4.0 environments," *2020 IEEE 24th International Enterprise Distributed Object Computing Workshop (EDOCW)*, 2020, pp. 91–94. DOI: 10.1109/EDOCW49879.2020.00025

12. M. -Q. Tran, M. Elsisi, K. Mahmoud, M. -K. Liu, M. Lehtonen, and M. M. F. Darwish, "Experimental setup for online fault diagnosis of induction machines via promising iot and machine learning: Towards Industry 4.0 empowerment," *IEEE Access*, vol. 9, pp. 115429–115441, 2021. DOI: 10.1109/ACCESS.2021.3105297

13. E. Balamurugan, L. R. Flaih, D. Yuvaraj, K. Sangeetha, A. Jayanthiladevi, and T. S. Kumar, "Use case of artificial intelligence in machine learning manufacturing 4.0," *2019 International Conference on Computational Intelligence and Knowledge Economy (ICCIKE)*, 2019, pp. 656–659. DOI: 10.1109/ICCIKE47802.2019.9004327

14. K. S. Kiangala and Z. Wang, "An adaptive framework for configuration of parameters in an Industry 4.0 manufacturing SCADA system by merging machine learning techniques," *2020 International Conference on Artificial Intelligence, Big Data, Computing and Data Communication Systems (icABCD)*, 2020, pp. 1–6. DOI: 10.1109/icABCD49160.2020.9183818

15. F. S. Cebeloglu and M. Karakose, "Comparative analysis of cyber security approaches using machine learning in Industry 4.0," *2020 IEEE International Symposium on Systems Engineering (ISSE)*, 2020, pp. 1–5. DOI: 10.1109/ISSE49799.2020.9272237

16. A. Dey, "Deep IDS: A deep learning approach for intrusion detection based on IDS 2018," *2020 2nd International Conference on Sustainable Technologies for Industry 4.0 (STI)*, 2020, pp. 1–5. DOI: 10.1109/STI50764.2020.9350411x

17. K. R. Thoorpu and N. Prafulla, "Sequential dtc vector embedding using deep neural networks for Industry 4.0," *2020 IEEE 7th International Conference on Industrial Engineering and Applications (ICIEA)*, 2020, pp. 912–915. DOI: 10.1109/ICIEA49774.2020.9102090

18. R. Ozdemir and M. Koc, "A quality control application on a smart factory prototype using deep learning methods," *2019 IEEE 14th International Conference on Computer Sciences and Information Technologies (CSIT)*, 2019, pp. 46–49. DOI: 10.1109/STC-CSIT.2019.8929734

19. M. Miškuf and I. Zolotová, "Comparison between multi-class classifiers and deep learning with a focus on industry 4.0," *2016 Cybernetics & Informatics (K&I)*, 2016, pp. 1–5. DOI: 10.1109/CYBERI.2016.7438633

20. H. Subakti and J. -R. Jiang, "Indoor augmented reality using deep learning for Industry 4.0 smart factories," *2018 IEEE 42nd Annual Computer Software and Applications Conference (COMPSAC)*, 2018, pp. 63–68. DOI: 10.1109/COMPSAC.2018.10204

21. H. Gao, Y. Peng, K. Jia, Z. Wen, and H. Li, "Cyber-physical systems testbed based on cloud computing and software defined network," *2015 International Conference on Intelligent Information Hiding and Multimedia Signal Processing (IIH-MSP)*, 2015, pp. 337–340. DOI: 10.1109/IIH-MSP.2015.50

22. W. Matsuda, M. Fujimoto, T. Aoyama, and T. Mitsunaga, "cyber security risk assessment on industry 4.0 using ICS testbed with AI and Cloud," *2019 IEEE Conference on Application, Information and Network Security (AINS)*, 2019, pp. 54–59. DOI: 10.1109/AINS47559.2019.8968698

23. N. Jazdi, "Cyber-physical systems in the context of Industry 4.0," *2014 IEEE International Conference on Automation, Quality and Testing, Robotics*, 2014, pp. 1–4. DOI: 10.1109/AQTR.2014.6857843

24. N. Moustafa, E. Adi, B. Turnbull, and J. Hu, "A new threat intelligence scheme for safeguarding industry 4.0 systems," *IEEE Access*, vol. 6, pp. 32910–32924, 2018. DOI: 10.1109/ACCESS.2018.2844794

25. Z. Rahman, I. Khalil, X. Yi, and M. Atiquzzaman, "Blockchain-based security framework for a critical Industry 4.0 Cyber-Physical system," *IEEE Communications Magazine*, vol. 59, no. 5, pp. 128–134, May 2021 DOI: 10.1109/MCOM.001.2000679

26. D. Lukač, "The fourth ICT-based industrial revolution "Industry 4.0"—HMI and the case of CAE/CAD innovation with EPLAN P8," *2015 23rd Telecommunications Forum Telfor (TELFOR)*, 2015, pp. 835–838. DOI: 10.1109/TELFOR.2015.7377595

27. S. Das, L. Ghosh, and S. Saha, "Analyzing gaming effects on cognitive load using artificial intelligent tools," *2020 IEEE International Conference on Electronics, Computing and Communication Technologies (CONECCT)*, 2020. DOI: 10.1109/CONECCT50063.2020.9198662

28. S. Das and A. Bhattacharya, "ECG assess heartbeat rate, Classifying using BPNN while watching movie and send movie rating through Telegram." In Tavares J.M.R.S., Chakrabarti S., Bhattacharya A., and Ghatak S. (eds) *Emerging Technologies in Data Mining and Information Security*. Lecture Notes in Networks and Systems, vol. 164. Springer, Singapore, 2021. https://doi.org/10.1007/978-981-15-9774-9_43.

29. Dnyandip Bhamare, Pranaynil Saikia, Manish Rathod, Dibakar Rakshit, and Jyotirmay Banerjee, "A machine learning and deep learning-based approach to predict the thermal performance of phase change material integrated building envelope," *Building and Environment*, vol. 199, p. 107927, 2021. 10.1016/j.buildenv.2021.107927.

30. Tianhong Mu, Fen Wang, Xiufeng Wang, and Hongjie Luo, "Research on ancient ceramic identification by artificial intelligence," *Ceramics International*, vol. 45, 2019. DOI: 10.1016/j.ceramint.2019.06.003

31. N. R. Mavani, J. M. Ali, S. Othman et al., "Application of artificial intelligence in food industry—A Guideline," *Food Engineering Reviews*, vol. 14, 2021. DOI: 10.1007/s12393-021-09290-z

12

Triangulation-Augmented AI-Algorithm for Achieving Intelligent Flight Stabilizing Performance Capabilities

Bibhorr

IUBH International University, Bad Honnef, Germany

Chirag R. Anand

Harvard University, Cambridge, Massachusetts, United States

CONTENTS

12.1 List of Symbols

Symbol	Denotation
Δk	Angular attitude change
X	Miscellaneous force
T	Thrust force
M	Resultant moment
m_i	Moment of impetus

DOI: 10.1201/9781003241348-12

m_r	Moment of restoration
l_N	Neutrality limit
l_p	Load on push actions
l_m	Load on mathematical actions
I	Orientation array/attitude array
x, y, z	Coordinate data
T	Time-stamp array
t_x, t_y, t_z	Time point data scores
$(I \cup B) \cup (T \cup A)$	Orientation modeling array
x', y', z'	Modified mathematical surging data
a, b, c	Revised mathematical data
F	Force array
M	Moment array
D	Angular deflection array
α, β, γ	Angles of deflection
R	Restoration force modeling
L	Total time limit of $f(x,t)$
k	Accumulation of max. time
Z	Total time limit of $f(y,t)$
N	Total time limit of $f(z,t)$
T'	Time matrix
V	Final deflection array
e	Encoding cycle
$f_i(t)$	Signal duration
t_i, T_i	Time instants
$\tilde{x}$	Continuous time reconstruction of input signal y

12.2 Background

Avionics systems are becoming more reliant on software for functioning, with related software encompassing a broad number of application fields. These include flight controls for airplanes and spacecraft, weapons management, defense mission program computation, command and control, sensor management and processing, surveillance, telemetry, and other functions. Flight control systems have progressed from entirely mechanical through hydromechanical to digital stability/control augmentation systems [1]. Flight control systems are recognized in a variety of forms, ranging from most simple to the most complicated, and improvements to these systems are continually being devised to make airplanes safer and easier to operate. Depending on the kind of aircraft flown, the flight control systems and characteristics might differ substantially. Primary and secondary systems are used in aircraft flight controls. The primary control systems consist of the elevator, ailerons, and rudder which are essential to operate an aircraft safely during the most duration of flight phase. The secondary control system consists of flaps, trim systems, and spoilers which increase the performance characteristics of the aircraft or also help relieve the pilot of excessive control forces.

The stability of an aircraft is an imperative prerequisite feature which is regardfully foresighted by aerospace engineers during the manufacturing, designing, research, and development phases. Early aircraft designs relied heavily on subtlety of flight stability and control. The introduction of automatic controls led to the accelerated development of commercial, cargo, and military aircrafts in the modern years. The extent to which this stability comes off may differ as per the stick-free and stick-fixed conditions. Apart from environmental disturbances, in some cases, powerplants can also shoot instability on an aircraft.

An aircraft is considered stable if it tends to return to its precursory function-centric fundamental equilibrium orientation after being displaced as a perturbing anomaly due to a disturbance caused by the impetus. If an aircraft fails to accomplish this tendency, then it is deemed unstable and thus critically unsafe. Therefore, it becomes clear that the response of an aircraft originates purely as a consequence of its deep-rooted built-in design criterion. The disturbance in the aircraft is caused either due to the environmental conditions or pilot's in-flight control actions during flight operation. The environmental disturbances include air turbulence, wind gusts, thunderstorms, heavy rainfall, gradients, etc.

At certain flight conditions, an airplane may bear stable performance, but in other instances, the chances are that it may not be as much stable as it is operated earlier, or it may bear complete instability. It means that if an airplane is stable during straight and level flight, it may become unstable during inverted flying regime, and the process is communicated vice versa. Some aircrafts, such as training aircrafts, are built to be precisely stable but vehicles like fighter jets and other military aircrafts tend to be very unstable. Some military combat aircrafts are purposefully designed to be inherently unstable, in order, to enhance agility and subsequently reduce the drag. These vehicles with lack of aerodynamic stability shoot up extremely hazardous consequential scenarios and hence are made to be piloted by highly experienced pilots. This, thus, necessitates the use of an advanced automatic artificial stabilization system that is completely reliable and efficient. Inherent instability is not accorded by aerospace engineers in civil aircrafts due to the potentially catastrophic effects of system failure. But with appropriate safety measures, it is possible to turn down the level of stability relative to previous versions of designs, which builds up benefits in the form of reduced drag. During the analysis of vehicle's stability in-flight dynamics, short- and medium-time responses of vehicle attitude and velocity are considered. Stability takes into account the vehicle's response to aberration for maintaining a smooth and safe flight. When an aircraft with one wing dipping toward Earth and the other pointing toward sky is flown, it is common for it to return to the initial position when disturbed. Such an aircraft is highlighted as being out of trim, it is not deemed unstable. There is an intermediate condition defined between stability and instability conditions of an aircraft. This is the condition followed by an aircraft, when disturbed, in which it tends to neither return to its equilibrium orientation nor it further delineates from the initial position and tends to retain the newer position. This intermediary neutral stability is an essentially preferred feature in aircrafts in some conditions.

When considering the issue of air traffic management, stability requirements become more complex. Aircrafts are expected to follow a reference trajectory, but aircrafts must be kept apart from each other to avoid close-ups and mid-air collisions [2]. Unstable control systems represent a potential hazard and are therefore the most important issue in stability control systems and engineering [3]. If an aircraft maintains a stable uniform flight highlighting angular attitude change Δk, then both the miscellaneous force x other than thrust T and the resultant moment m around the center of gravity must equal zero. An airplane that meets this requirement is said to be in a state of equilibrium. This criterion is mathematically defined as:

$$\Delta k = f(x, m) = 0 \tag{12.1}$$

Based upon the aspects, stability is fundamentally classified into static and dynamic stability. Static stability indicates the initial tendency of the vehicle's response to perturbation that leads to equilibrium recovery. Dynamic stability is associated with the time-mapped blueprint of the vehicle's movements after being disturbed from the equilibrium point. The vehicle that is statically stable can be deemed dynamically unstable. Therefore, static stability does not guarantee any kind of dynamic stability.

Figure 12.1 illustrates the typical behavior response the vehicle can show up post the disturbance caused in the trajectory due to an impetus. The outcome of vehicle's behavior demotes a reaction to the action. The action, which in this case is movement caused due to an impetus, may lead to statical stability, neutrality, or instability in behavior. On the contrary, Figure 12.2 illustrates the dynamic stability response condition which highlights the typical behavioral response to an impetus leading the vehicle to maintain an equilibrium state.

The authors mathematically define the behavior scenario of static stability as:

$$m_i = m_r \,|\, m_r < l_N \text{ (statically stable condition)} \tag{12.2}$$

$$m_i = l_N \text{ (statically neutral condition)} \tag{12.3}$$

$$m_i > l_N \text{ (statically unstable condition)} \tag{12.4}$$

where m_i denotes moment of impetus, m_r signifies moment of restoration, l_N is the neutrality limit.

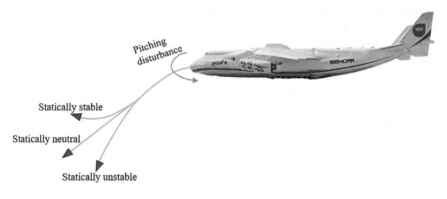

FIGURE 12.1
An illustration of response to a stimulus that can reflect statical stability, neutrality, or unstability in behavior.

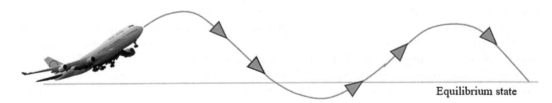

FIGURE 12.2
An illustration of dynamic stability response condition.

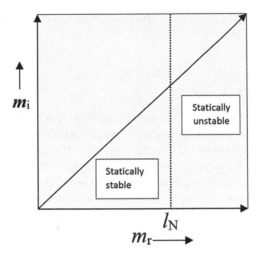

FIGURE 12.3
A graphical plot for moment of impetus m_i vs. moment of restoration m_r.

The neutrality limit is the restoration moment limit crossing which the vehicle fails to signal any recovery and is deemed critically unsafe. In Figure 12.3 (shown below), the relationships among m_i, m_r, and l_N are demonstrated in a graphical visualization. Along x-axis runs restoration moment and along y-axis runs moment of impetus.

12.3 Triangulation-Augmented Artificial Intelligence (AI)-Algorithm

12.3.1 The Founding Concept

Triangulation is a multidimensional approach for acquiring precise measurements in engineering applications that comprises a set of computational sub-procedural geometrically matched patterns. It is a basic procedure for analyzing computational geometric problems in which complex geometric figurines are first broken down into the simplest forms possible, such as tetrahedrons and triangles, depending on the type of problem, and then analyzed further to find the desired engineering solutions.

Triangulation is a crucial procedure in computer geometry and graphics. Triangulation enables three-dimensional objects to be represented by a series of points. This procedure is critical for the speed, quality, and resolution of the shown items. Polygon triangulation is used in geo-information systems and in the process of digital terrain modeling [4]. Triangulation is the technique of calculating the 3D location of a point using two photographs in which that point is visible. This procedure requires the intersection of two known lines in the space. If, however, this intersection does not occur due the presence of noise, the best approximation must be estimated [5].

Triangulation can be understood as a subdivision of tessellation into a number of connected but non-overlapping triangles. The triangles within a triangulation are established through points in the domain Ω. During the construction of triangulations, the collection of points is represented as [6]:

$$P = \{p_i\}, i = 1, \ldots, N,$$

A wide range of optical techniques for the deposition development monitoring depends on the triangulation rule. Its conventional style execution takes advantage of a shifted laser examining the objective surface, with a picture sensor utilized for identifying the test spot position [7]. Triangulation surveying method is based upon the trigonometric premise that if one side and the angles of a triangle are known, then the unknown sides can be easily evaluated. Triangulation methodology in conventional surveying involves measuring of base which is a line forming a side of one of the triangles, measuring angles of each triangle and computation of distances between two points through sequential triangles in regular order. The triangulations, thus formed, are categorized on the basis of precision of measurement, viz. first order, second order, and third order [8]. There are several systems that use and heavily rely upon triangulated network computation—some of these systems primarily include Global Navigation Satellite, Position Resection, and RTL Systems. Triangulations are used for various simulations and particularly in aerodynamics where triangulations can be used to fabricate an error-free lattice structure during mesh generation criterion. Triangulation techniques when used in complex and advanced scenarios involve effective integration in trajectory identification, trajectory transitions, and aerodynamics performance parametrization and efficient space traffic management. Today, with so much changing each day and new concepts and AI-based solutions challenging conventional and outdated methods, the scope of triangulation implementation in space traffic management is enormous in terms of maintaining a secure grip in future technology development. Utilizing the potential of triangulation concept, this work incepts an AI-algorithm for achieving flight stabilizing performance capabilities.

12.3.2 The Inception

Kisabo et al. (2012) in [9] inquired into the design, modeling, and analysis of two autopilots: a fuzzy proportional-integral-derivative (PID) controller and its hybrid with a PID controller for the control of an aircraft's pitch plane dynamics. The Mamdani-type fuzzy inference system was used for the fuzzy inference system (FIS) in the fuzzy logic controller design in their work. The hybrid fuzzy PID controller criterion was detailed by S.N. Deepa et al. (2013) in [10] to increase the performance of an aircraft's longitudinal control. Similar works using fuzzy algorithms are on the slow rise but the core problem solving in ultra-critical real-time situations remains blurry on the practical scenario. Due to the lack of core engineering work done to employ AI in this stream of aerospace domain, the authors considered key essentiality working on this problem. Due to the limitations of fuzzy logics viz., predictions not always being accurate, ambiguity involved in certain situations, and slow run time, authors found it apt to look for new models and algorithms leading to the development and inception of this work. The incepted algorithm is based on the triangulation logic where the concept allows for the multiple logical values of a variable or problem spanning between 0 and 2π.

12.4 AI-Algorithm Schema

As evident in Figure 12.4, the schema of AI-backed algorithm is determinatively incepted through a set of disparate actions peculiarized broadly in two traces—push actions denoted by p and mathematical actions denoted by m. Although all the process actions (push and mathematical) involve mathematical modeling criteria, to some extent, they are established disjointly because of their dominant capacity to achieve the desired action based on factors like communication, magnitude of mathematical modeling, mechatronic processes, etc. The actions p and m are time-bound because the criticality of the attitude perturbation is the key feature of the algorithm design. The purpose of classifying action sequence in terms of m and p codes, given as below in Figure 12.4, is to strategize the parametric success distribution of overall functioning of the AI-algorithm. The addition, subtraction or replacement of the actions could be accommodated well through this plan post strategizing the new experimental requirements. In order to increase the success rate of the AI-algorithm, changes lie in switching these codes in different combinations. Thus, this plan necessitates the algorithm's success rate alteration.

For a unit action load l, the algorithm consists distribution of 0.6 fractions on purely mathematical actions and exact distribution of 0.4 fractions on push actions, which are partially mathematical in nature. The push actions as the name suggests are the actions that advance the processes by performing the dispensed tasks in the limited frame of time. These tasks vary from time stamping to actuating controls of the vehicle, as they denote tasks of communication, mechatronic processes, digital handling, hybrid identification, minor data filtering, and other similar tasks. The mathematical actions involve the tasks that require complicated computations, simulations, validations, pattern analysis, prediction judgments, crucial decisions, etc. The unit action load (UALD) distribution for this model is given as:

$$l = l_p + l_m \tag{12.5}$$

$$l = 0.6 + 0.4 = 1 \tag{12.6}$$

where l_p denotes the load on push actions and l_m denotes the load on mathematical actions.

The attitude identification I is accomplished by parallelly communicating signals directly from the attitude indicators or gyroscopic devices and systems as and when displayed on the avionics. The inputs identified in the first process are based on the orientation through coordinate data as:

$$I = \{x, y, z\} \tag{12.7}$$

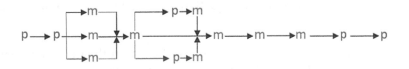

FIGURE 12.4
Action sequence plan of AI-algorithm for attitude stabilization.

where x, y, and z are the coordinates in their respective directions. After the completion of this action, time stamping is initiated whereby each processed identification data score is digitally time-stamped. The time-stamp array T is represented as:

$$T = \{t_x, t_y, t_z\} \tag{12.8}$$

where t_x, t_y, and t_z are the time point data scores corresponding to x, y, and z coordinates, respectively.

The primary orientation modeling comes into picture post the end of the stamping action. This action being the representative of mathematically dominant processes transforms the data into a new code that can be further send for computations and simulations. The orientation modeling array $(I \cup B) \cup (T \cup A)$ is the result of an agglutination process and is mathematically given as:

$$(I \cup B) \cup (T \cup A) = \{x, y, z, t_x, t_y, t_z, x', y', z', a, b, c\} \tag{12.9}$$

where

$$I = \{x, y, z\};\ B = \{x', y', z'\};\ A = \{a, b, c\}\ \text{and}\ T = \{t_x, t_y, t_z\} \tag{12.10}$$

Here, B is the modified mathematical surging data and A is the revised mathematical data. Most of the times, the following equations are satisfied:

$$x' = a;\ y' = b;\ z' = c \tag{12.11}$$

However, this is not always the case.

For simplicity, the scenario from Equations (12.2) to (12.11) is recalled briefly as:

- Equations (12.2), (12.3), and (12.4) describe the static stability scenario in terms of moment of impetus, moment of restoration, and neutrality limit.
- Equations (12.5) and (12.6) describe load distribution scenario on push and mathematical actions.
- Equation (12.7) denotes attitude array obtained as a result of gyroscopic signal received.
- Equation (12.8) describes time-stamp array obtained as a result of time-stamping phase initiation.
- Equation (12.9) describes orientation modeling array.
- Equation (12.10) describes modified mathematical surging data and revised mathematical data.
- Equation (12.11) further describes equations for modified mathematical surging data scores.

The estimation of forces and moments is accomplished from the metrology data and the time-stamped data received. This estimation action is given as:

$$F \cup M = \{p_x, p_y, p_z, p_R, m_x, m_y, m_z, m_R\} \tag{12.12}$$

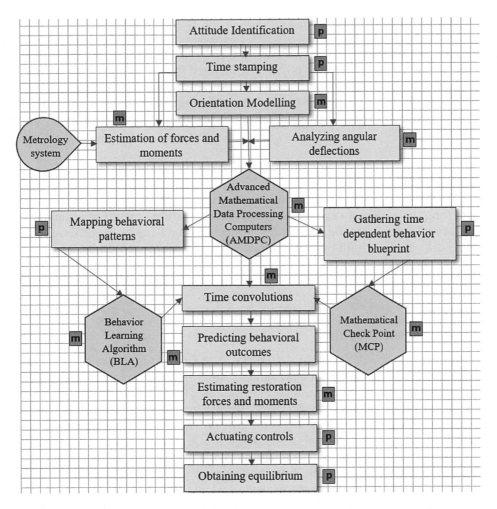

FIGURE 12.5
Design architecture of AI-algorithm for attitude stabilization.

where p with subscripts x, y, z denotes the forces and m with same subscripts denotes the moments along their respective axial directions. p_R and m_R are the resultant force and moment, respectively. Here, F is the force array and M is the moment array given as:

$$F = \{p_x, p_y, p_z, p_R\} \tag{12.13}$$

$$M = \{m_x, m_y, m_z, m_R\} \tag{12.14}$$

Angular deflections D are analyzed parallelly along this process and is extensionally written as:

$$D = \{\alpha, \beta, \gamma\} \tag{12.15}$$

where α, β, γ are the angles of deflection along the x, y, and z directions, respectively.

The information is then channelized to advanced mathematical data processing computers (AMDPC) where the data are filtered, processed, analyzed, computed, simulated, and distributed further for mapping behavioral patterns and gathering time-dependent behavior blueprint. The information is then channelized for establishing firm algorithm for behavior learning and mathematical checks to ultimately produce time convolution data sets. These convolution data sets are helpful in predicting the behavioral outcomes for the estimation of restoration forces and moments. The restoration force modeling R is mathematically given as:

$$R = \{r_x, r_y, r_z, r_R\, n_x, n_y, n_{z,}\, n_R\} \qquad (12.16)$$

where r with subscripts x, y, z denotes the forces and n with the same subscripts denotes moments along their respective axial directions. Here, r_R and n_R are the resultant force and moment, respectively. After the estimation of moments and forces, controls are actuated accordingly and finally equilibrium is attained.

12.5 Attitude Identification

Attitude indicator is one of the flight-system instruments that indicates the orientation of the spacecraft providing information on vehicle's angles in relation to the natural horizon in order to reorient the spacecraft and counter spatial disorientation. Attitude indicator is essentially a gyroscopic flight instrument with miniature of the vehicle and the line of horizon on display. The miniature vehicle on the display mimics the real-time orientation of the vehicle to the actual horizon. For presenting attitude information, it can be designed in either a moving horizon format or the moving vehicle format. The limits of the pitch-and-bank angles are usually dependent upon the design of the model. The attitude indicator is one of the most definitive flight instruments and provides the most accurate information of the representational vehicle in real time on the instrument panel. The display that is designed based on moving horizon format has the vehicle indicator as a stable element while the artificial horizon as the moving element according to the real-time scenario and positioning of the natural horizon with respect to the vehicle. In the moving horizon format, banking of the vehicle is in the direction opposite to the movement of the artificial horizon, that is, banking to the left is indicated by movement of the artificial horizon to the right while pitching of the vehicle is indicated by upward and downward movements of the horizon line. The moving vehicle format design, on the other hand, displays the vehicle representation in the center and an artificial horizon line where the bank angle is displayed by the movement of the vehicle representation with the artificial horizon kept in a steady horizontal position in reference to the instrument panel while the pitch angle is indicated by the upward and downward movement of the artificial horizon line, similar to the other format. The gyro must remain vertically erect while the vehicle rolls and pitches around it for the attitude indicator to function effectively (Figure 12.6).

Although least friction is encountered in such devices, even a minimal amount of friction acts as a constraint on the gyro, causing precession and tilting. An erection mechanism inside the instrument casing provides a force whenever the gyro tilts from its upright position to reduce this tilting. This force serves to bring the spinning wheel back to its upright

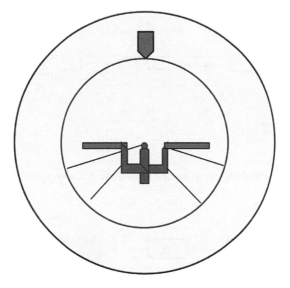

FIGURE 12.6
An illustration of attitude indicator.

position. The caging mechanism that locks the gyro in case of vehicle maneuvering exceeding the limits of the instruments is absent in newer instruments due to absence of any tumble limits. The gyro is not erect at the ignition of the engine. A self-erecting mechanism inside the instrument, triggered by gravity, applies a processing force, which causes the gyro to ascend to its vertical position. Such gyroscopic instruments associated with attitude indicator and other flight-system instruments have been replaced with advanced attitude and head referencing systems, whose functions remain similar to the gyroscopic systems; with the initial heading indication, the attitude and head referencing systems can determine both the attitude and magnetic heading. Stabilization and control of the vehicle in line with spatial orientation can be achieved through pitch-and-bank control. The angular relationship between the vehicle's longitudinal axis and the real horizon is controlled by pitch attitude control. The attitude indicator provides a clear and immediate indication of the pitch attitude. For any pitch attitude required, the controls are utilized to place the vehicle representation on display, in reference to the horizon bar or horizon line. Although the attitude indicator is the major attitude reference, the concept of primary and supporting instruments does not diminish the importance of any specific flight instrument in setting and sustaining pitch-and-bank attitudes when it is available. It is the only instrument that accurately depicts the actual flight attitude in real time.

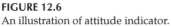

12.6 Time Stamping

After attitude identification, the time-stamping phase comes in, where each data set is time-stamped for the further execution and management of artificial intelligent performance algorithm. The time-stamping phase forms the small constituent dealing within the whole AI-based attitude stabilization model. Figure 12.7 describes the time-stamping

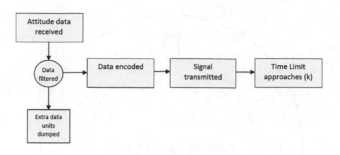

FIGURE 12.7
Flowchart describing the time-stamping phase in the AI-based stabilization algorithm.

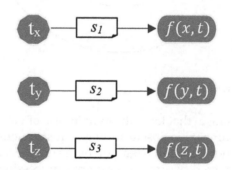

FIGURE 12.8
Time-stamped data sets shown as function of time and axially described coordinates.

phase which begins with the reception of attitude data and culminates when the time limit approaches k. This phase consists of data encoding process and data dumping. The extra data are deliberately leaked and dumped as the system picks only nominal data that is, in reality, necessary for further computations.

The time point data scores, as shown in Figure 12.8, notated as t_x, t_y and t_z are functions of time and axial coordinate data. These are mathematically written as:

$$tx = f(x,t) = \int \frac{1}{x} dt \qquad (12.17)$$

$$ty = f(y,t) = \int \frac{1}{y} dt \qquad (12.18)$$

$$tz = f(z,t) = \int \frac{1}{z} dt \qquad (12.19)$$

The total time limit of function $f(x,t)$ is L when time t approaches k. Here, k signifies the accumulation of maximum time allotted to the action. This is mathematically notated as:

$$\lim_{t \to k} f(x,t) = L \qquad (12.20)$$

The total time limit of function $f(y,t)$ is Z when time t approaches k. Again, k signifies the accumulation of maximum time allotted to the action. This is mathematically notated as:

$$\lim_{t \to k} f(y,t) = Z \tag{12.21}$$

Similarly, for (z, t), the equation is given as:

$$\lim_{t \to k} f(z,t) = N \tag{12.22}$$

The time matrix for each coordinate data is represented in primordial element, coded element, and final output element. This three-value time array has been represented as below:

$$T' = \begin{bmatrix} a \\ b \\ c \end{bmatrix} \tag{12.23}$$

where T' denotes the time matrix and a, b, and c are the primordial, coded, and final output elements, respectively.

12.7 Orientation Modeling

Orientation modeling begins with the reception of time-stamped data and ends after the emittance of computer readable data. As soon as the time-stamped data are received, data are re-encoded in suitable format, and then post the simulation process, minute elements are detected. These minute elemental details are permanently archived or retained for backup, and the duplicated readable data are emitted for the completion of the phase. This process is not typically time-bound; however, an upper preference time limit always helps in secure the fast processing (Figure 12.9).

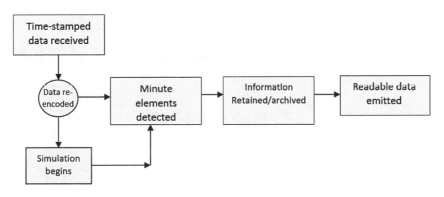

FIGURE 12.9
Flowchart depicting orientation modeling.

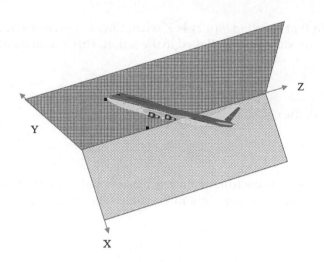

FIGURE 12.10
Orientation modeling in AI-based attitude stabilization algorithm.

The simulation of the altering coordinate pixels is rendered on a three-dimensionally digitally distended frame, and the magnitudes of positions are encoded in similar proportion for further sending data for accomplishing mathematical modeling. This is the pre-phase of estimating forces and moments. As shown in Figure 12.10, the data received are encoded in pixel or other format suitable for computers to compute and analyze the data. The readable data are output of this action, which is then facilitated further for force and moment estimation.

12.8 Angular Deflection and Moment Estimation

In this phase of algorithm action, the current position of the aircraft is digitally made to imbricate with the intended equilibrium position. This is done by computing the difference between the current coordinate pixel data and the intended equilibrium pixel elemental data. The more the mismatch, the more the deflection and more is the workload distributed throughout the algorithm. The deflections are analyzed along all the axial directions plus the resultant one (Figure 12.11).

The deflection data goes through number of filtering and coding mathematical procedures until the following final deflection array V is obtained:

$$V = \{q_\alpha, q_\beta, q_\gamma, q_x, q_y, q_z\} \tag{12.24}$$

Here,

$$q_\alpha = \frac{p_x}{\alpha}; \; q_\beta = \frac{p_y}{\beta}; \; q_\gamma = \frac{p}{\gamma} \tag{12.25}$$

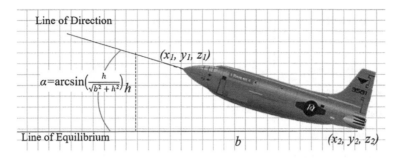

FIGURE 12.11
Angular deflection in AI-based attitude stabilization algorithm.

And also,

$$q_x = \frac{m_x}{x}; q_y = \frac{m_y}{y}; q_z = \frac{m}{z} \tag{12.26}$$

Recalling data from Equations (12.13) and (12.14), p with subscripts x, y, z denotes the forces and m with same subscripts denotes the moments along their respective axial directions. Also recalling data from Equation (12.15), α, β, γ are the angles of deflection along the x, y, and z directions, respectively.

12.9 Advanced Mathematical Data Processing Computers

As shown in Figure 12.12, the concept of AMDPC consists of three main phases, viz., data input phase, data processing phase, and data output phase. Data input phase involves actions such as data gathering/collection, data filtering, deliberate data leakage for increasing speeds, encoding and decoding process cycles, data communication, data transmission, and data archiving/storing. The next phase is the data processing phase consisting of simulations performance, data analyses, behavior analyses, execution of mathematical operations, detection of error indices, manipulation and correction of computational errors, estimation of data blueprints, and finally sending back unidentified data. The last phase of AMDPC highlights the classification of data, soring of data, encoding, decoding cycle, data presentation for further transmission, and data archiving for future references in case of system malfunction.

If the typical data received are in the form of array A, B, C, D, and F as:

$$A = \{x, y, z, k, l, m, u, v, y\} \tag{12.27}$$

$$B = \{x', y', z', k', l', m', u', v', y'\} \tag{12.28}$$

$$C = \{x_1, y_1, z_1, k_1, l_1\} \tag{12.29}$$

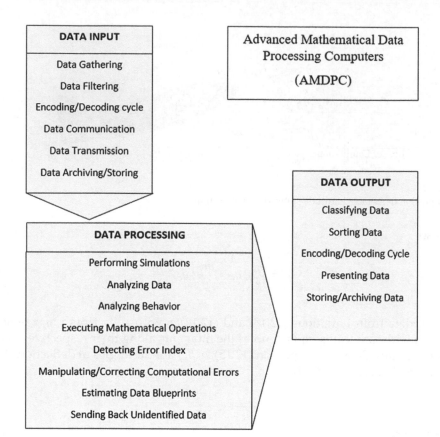

FIGURE 12.12
Flowchart depicting functions of advanced mathematical data processing computers.

$$D = \{x_2, y_2, z_2, k_2, l_2, m_2, u_2, v_2, y_2\} \tag{12.30}$$

$$E = \{x, y, k, l, u, x_1, y_1, u_2, v_2\} \tag{12.31}$$

Then, the typical criteria for filtering the data are performed by eliminating the repeated elements (such as in E) and also the elements of equivalence (C and D), such that only the following data sets are chosen for further processing:

$$A = \{x, y, z, k, l, m, u, v, y\} \tag{12.32}$$

$$B = \{x', y', z', k', l', m', u', v', y'\} \tag{12.33}$$

This type of filtering is beneficial because it helps increase the data processing speed and also helps prevent storage accumulation lag that may occur due to the excessive archival of unwanted data.

For the input data i, the encoding cycle e works based upon the equations as follows:

$$e = \int f(i) f(t - i) \, di \tag{12.34}$$

where t represents the encoding element.

The encoding can also be manipulated as:

$$e = f(t) \int f(i)\,di \tag{12.35}$$

where $f(t) = k't$ in which k' denotes the encoding constant that may depend upon the algorithm's random pick.

12.10 Behavior Learning Algorithm (BLA)

Bibhorr (2019) in [11] gave a triangulation-determining formula which has been utilized in establishing the resultant outcome for learning scenario in AI-algorithm. Behavior learning algorithm is based upon the triangulation modeling where the angular magnitude is the indicator of learning workload. The behavior learning algorithm adjusts itself based on the angular variation. The algorithm first receives the data array for the time interval t_1 and when further it receives data for time interval t_2, the algorithm computes the variation in the data in angular form. As shown in Figure 12.13, the algorithm computes the angular variation θ_1 as a function of variation between the data lines for two different time intervals t_1 and t_2.

The algorithm computes the angular variations for the given array of data and then initiates the prediction algorithm by averaging out the angular variations for the prediction of the data for the future time intervals. The algorithm functions as a continuous chain of predicting variations. The angle θ is a sum of functions of the vertical q, horizontal r, and the diagonal p of the triangulated lattices. This is mathematically given as:

$$\theta = \Sigma f_i(p,q,r) \tag{12.36}$$

where

$$f(p,q,r) = \frac{\pi}{2} - \frac{\pi/2\,(p+q-r)^2}{q^2 + \dfrac{3p}{2}(p+q-r)} \tag{12.37}$$

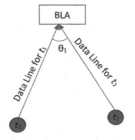

FIGURE 12.13
Depiction of the behavior learning algorithm.

12.11 Time Convolution

Signal processing employs convolution operations throughout predominantly for its effective representation of linear, time-invariant systems. For two signals x and y, the convolution, in discrete-time, is defined as [12]:

$$(x*y)(n) = \sum_{k=-\infty}^{\infty} x(k)y(n-k) \tag{12.38}$$

$$(x*y)(n) = \sum_{k=-\infty}^{\infty} y(k)x(n-k) \tag{12.39}$$

Basically, the convolution advocates the output of the system relying on the input signal. Linear time-invariant systems can be outlined by impulse responses. The input signal to the system can be computed by executing the sum of the unit impulses scaled and moved by the sort attribute of the discrete-time impulse function. Therefore, the linearity infers that it seems reasonable to evaluate the output signal and calculate it as the sum of the scaled and moved unit impulse responses. The convolution can be used to determine the output of a linear time-invariant system from the supplied input and the impulse response.

Given the two transforms $F(z)$ and $G(z)$ such that f and g are their inverses, then the inverse h, which is the product of f and g, is given as [13]:

$$H(z) = F(z)G(z) \tag{12.40}$$

The inverse h which is written as $f*g$ is the convolution of f and g. The classical convolution theory states that H is the transform of convolution h of f and g notated as:

$$h(t) = (f*g)(t) = \int_{0}^{t} f(t-s)g(s)ds \tag{12.41}$$

12.12 Restoration Moment Estimation

The hybridization of angular deflection and moment estimation phase and time convolution phase results in the estimation of restoration forces and moments. The optimal plotting of time convolutions helps achieve the required estimation criteria. The blueprint of the whole disturbance is observed in comprehensive step-by-step approach by the algorithm, and the equilibrium restoration operation is initiated following the reverse conditions of the stepwise blueprint data. If during this reversal operation, the algorithm detects an error, then the additional forces and moments are evaluated and considered, and control surfaces are actuated accordingly. This action leads to the restoration of equilibrium status of an aircraft which was found to be under the major disturbance. The restoration moment m_r can be defined as the reverse integration of the function of the final simulated behavioral blueprint data b. This is mathematically given as:

$$m_r = -\int f(b)\, db \tag{12.42}$$

where the minus sign on the right-hand side of Equation (12.42) denotes the reversal process.

12.13 Experimental Results

Due to the moment of impetus m_i, the vehicle suffers angular attitude change Δk. To stabilize the condition, the algorithm generates prediction of restoration angle b, which when taken into consideration, the algorithm manipulates controls yielding moment of restoration m_r to finally reinstate the vehicle through actual restored attitude angle Δj. This scenario under experimental test conditions for a prototype vehicle is shown below. In the data, w denotes the unmediated standard hash term for digitally simulated prototyping moment.

Δk (°)	m_i (w)	m_r (w)	b (°)	Δj (°)
1	0.66	0.64	0.9918	0.9
1.5	0.68	0.67	1.5016	1.5
2	0.78	0.79	2.0012	2
2.5	0.79	0.81	2.5061	2.5
3	0.81	0.83	3.011	3
3.5	0.85	0.84	3.4991	3.4
4	0.89	0.91	4.0128	4
4.5	0.93	0.95	4.5218	4.5
5	0.97	0.98	5.016	5
5.5	1.12	1.1	5.5	5.5
6	1.3	1.3001	6.0001	6
6.5	1.5	1.5004	6.51018	6.5
7	1.68	1.69	7.019	7
7.5	1.79	1.83	7.506	7.5
8	2	2.2	8.0199	8
8.5	2.17	2.3	8.512	8.5
9	2.34	2.5	9.009	9
9.5	2.44	2.41	9.4966	9.4
10	2.6	2.7	10.1210	10.1
10.5	2.88	2.89	10.5	10.5
11	3	3.12	11.1023	11.1
11.5	3.2	3.3	11.513	11.5
12	3.39	3.4	12.001	12
12.5	3.5	3.49	12.5	12.5

(Continued)

(Continued)

Δk (°)	m_i (w)	m_r (w)	b (°)	Δj (°)
13	3.7	3.5	12.993	12.9
13.5	3.9	3.77	13.4986	13.4
14	4.8	4.99	14.004	14
14.5	5.1	5	14.49	14.4
15	5.3	5.37	15.102	15.1

FIGURE 12.14
Prototype of vehicle upon which simulation data have been gathered.

12.14 Conclusion

The newly formulated AI-augmented performance algorithm for attitude stabilization is presented and demonstrated for applications in aerospace engineering. The research describes the novel design formulation of triangulation-augmented AI-algorithm for stabilizing the disturbed trajectories and attitudes of the space vehicles. The algorithm works by channelizing the data through various systems and computers for rendering computations and simulations. The algorithm design helps attain equilibrium by automatically actuating the control surfaces with the application of forces and moments that are estimated for its restoration condition by the algorithm itself. The research stresses on the critical mathematical modeling within the algorithm design crucial for maintaining the coherence throughout all the phases of actions. The algorithm might prove successful for future considerations in the practical engineering environment as it highlights the intelligent phenomena which ultimately helps distribute the pilot workload. Although this chapter deals with the application of the formulated algorithm in cyber-physical avionics systems, it could potentially be used in other engineering applications too.

References

1. D. C. Sharp, A. E. Bell, J. J. Gold, K. W. Gibbar, D. W. Gvillo, V. M. Knight, K. P. Murphy, W. C. Roll, R. G. Sampigethaya, V. Santhanam, S. P. Weismuller (2010) "Challenges and solutions for embedded and networked aerospace software systems," *Proceedings of the IEEE*, 98(4), pp. 621–634.
2. Z. H. Mau, E. Feron, K. Bilimoria (2001) "Stability and performance of intersecting aircraft flows under decentralized conflict avoidance rules," *IEEE Transactions on Intelligent Transportation Systems*, 2(2), pp.101–109.
3. J. G. Juang, L. H. Chien, F. Lin (2011) " Automatic landing control system design using adaptive neural network and its hardware realization," *IEEE Systems Journal*, 5(2), 266–277.
4. M. Saračević, Š. Plojović, & S Bušatlić (2020) IoT Application for Smart Cities Data Storage and Processing Based on Triangulation Method. *Internet of Things*. Springer, Cham.
5. I. Vite-Silva, N. Cruz-Cortés, G. Toscano-Pulido, L. G. de la Fraga (2007) Optimal Triangulation in 3D Computer Vision Using a Multi-objective Evolutionary Algorithm. In: Giacobini M. (eds) *Applications of Evolutionary Computing*. EvoWorkshops 2007. Lecture Notes in Computer Science, vol 4448. Springer, Berlin, Heidelberg.
6. Ø. Hjelle, M. Dæhlen (2006) *Triangulations and Applications*. Springer Science & Business Media, Berlin, Heidelberg.
7. S. Donadello, M. Motta, A. G. Demir, B. Previtali (2019) "Monitoring of laser metal deposition height by means of coaxial laser triangulation," *Optics and Lasers in Engineering*, 112, 136–144.
8. U.S. Coast and Geodetic Survey, Triangulation, U.S. Government Printing Office.
9. A. B. Kisabo, F. A. Agboola, C. A. Osheku, M. A. L. Adetoro, A. A. Funmilayo (2012) "Pitch Control of an Aircraft Using Artificial Intelligence," *Journal ofScientific Research & Reports*, 1, 1–16.
10. S. N. Deepa, G. Sudha (2013) "Longitudinal Control of an Aircraft Using Artificial Intelligence," *International Journal of Engineering & Technology*, 5(6), 4752–4760.
11. Bibhorr (2019) "Analytical Study of Measurement Methods in Engineering Applications Using Triangulation Techniques," *National Seminar on Use of Scientific and Technical Terminology in Science and Technology, Commission for Scientific & Technical Terminology, MHRD*.
12. H. Liu, J. Kotker, H. Lei, B. Ayazifar, *Discrete-Time Sytems and Convolution*, University of California, Berkeley.
13. M. Bohner, G. S. Guseinov (2007). "The Convolution on Time Scales," *Abstract and Applied Analysis*, 2007, 1–24.

13

Forecasting-based Authentication Schemes for Network Resource Management in Vehicular Communication Network

Vartika Agarwal and Sachin Sharma

Graphic Era (Deemed to be University), Dehradun, India

CONTENTS

13.1 Introduction

Various types of devices such as sensors, transmitters, and receivers are used for communication between vehicles. These devices enhance the experience of driving. If we talk about safety point of view, vehicles informing other drivers about any emergency can prevent the accident entirely. We have various forecasting-based authentication schemes, which help us to predict about resources and beacons in advance. Prediction-based authentication (PBA) schemes work for instant verification. With the help of schemes, we can predict the vehicle movement between two beacons. PBA is safe and secure and analyzes the packet loss effect. PBA record message is in the form of code. TESLA scheme is based on cryptography. If message is not delivered to another vehicle within a specified amount of time, then the message will not be sent again. Here, we can predict that if the message will not be delivered within a specific amount of time. There must be chances of security attack. Elliptic curve digital signature technique (ECDSA) is a digital signature, which is attached with message. With this signature, we can predict the beacons in advance and maintain the confidentiality of message. Prediction of location scheme can predict the location of vehicle in advance so that we can analyze the arrival of next upcoming beacon.

DOI: 10.1201/9781003241348-13

13.2 Literature Review

Many related studies have been reported on the authentication schemes for vehicular communication network (VCN) resource management. To achieve message authentication and security, we reviewed about forecasting-based authentication scheme for VCN resource management. In 2007, Vinod Namboodiri introduced prediction-based routing protocol (PBR). This protocol was extensively evaluated for changing vehicle density. PBR offers significant reduction in route failure and leads to high packet delivery [1]. In 2010, Vaishali discussed about the communication process between two or more automobiles. She tried to compare the performance and drawback of different simulators, which are used for establishing the communication between vehicles [2]. In 2015, Chen Lyu presented an effective scheme for authentication, which helps to protect information from different kinds of security attack. PBA stores shortened authentication code of message signature. This scheme works well for predicting beacons in advance [3]. In 2017, Abdullah Al Mamun presented a review of load-forecasting schemes. This scheme is used to control various operations such as dispatch and full allocation. This chapter focuses on hybrid method using machine learning [4]. In 2017, Joerg Evermann focused on event prediction in business process with recurrent neural network. [5]. In 2018, Harmanjot Kaur described the use of clustering-based routing approach and mainly focused on the merits and demerits of each technique [6]. In 2019, Mohammad Masdari presented a workload prediction scheme to improve resource management in cloud computing. They introduced load-forecasting schemes to recognize more realistic and complex request patterns, which may happen in real life [7]. In 2020, Mustafa Maad Hamdi presented an overview of the concept of vehicular ad-hoc networks, application, and features. They described various tools such as–GrooveSim, NHTSA, FLEETNET, or many other for solving the vehicular ad-hoc network simulation problems [8]. In 2020, Raj. K. Jaiswal introduced Kalman Filter for position-based routing protocol. These predictions were implemented using C++ library, and it has a great efficiency as compared to other protocols [9]. In 2020, Vartika et al. highlight various technology such as Bluetooth, Wi-Fi for V2V communication [10]. In 2020, Vartika reviewed about use of internet of things in transportation management system. They highlight traffic management module for managing traffic [11]. In 2020, Muhammad Mansoor compared two artificial neural network techniques, which is used for load forecasting [12]. In 2021, Vartika et al. presented a review chapter on network scheduling technique for scheduling a task. Such techniques are very efficient and effective and provide an overview of whole scheduling process [13]. In 2021, Vartika et al. investigated the deep learning techniques to improve radio resource management in VCN. The techniques of load forecasting and bidirectional long-short term memory (LSTM) will be discussed. Bi-LSTM network can be trained by input variables. The prediction model for financial data may use LSTM network for prediction of future data with better accuracy and superior performance over other kind of prediction model. An LSTM-based forecasting method has the power to capture any pattern in time series data. Its performance is better in comparison of other tested methods. A misbehavior detection system can detect and prevent internal attacks. In addition, Dempster-Shafer and beta distribution–based theory model for improving the accuracy of system can be used. This scheme works well in comparison to other misbehavior identification scheme [14]. In 2020, Hui Liu introduced multiobjective optimization-related theories. In this chapter, comprehensive review on this technology is done. They discuss the advantages and disadvantages of this technology and also highlight the research gaps for further research. A short-term load and price forecasting has used mutual information and random forest and recursive feature elimination technique. UMASS data set for price and load prediction can be used. This technique beats other tested methods in terms of accuracy and performance [15]. Zongjian He proposed a software-defined network–based architecture for communication between vehicles. This architecture increases the speed of communication and provides better throughput [16]. P. Deepalaksmi proposed an elliptic curve digital signature technique for the detection of malicious nodes. This scheme improves network performances and provides higher throughput [17]. Chia-Hui Liu proposed a user authentication scheme that monitors blood pressure, heart rate, and body temperature. They used wireless sensor network to store authenticated data. [18]. In 2021, Azeem Irshad designed a secure authentication scheme for smart grid–based demand resource management. This scheme works well for solving privacy as well as security issues [19]. In 2021, Sajid Hussain proposes elliptic curve cryptography–based authentication schemes for establishing communication between vehicles. This technique offers more secure communication between vehicles in comparison of other techniques [20]. In 2021, Dipanwita Sadhukhan proposed a user-based authentication scheme which runs on smart devices and maintains the confidentiality, security, and integrity of communication. AVISPA simulation tools confirm the robustness of this scheme [21]. In 2020, Mohd Shariq proposed a radio frequency identification scheme. The Scyther simulation tool verifies the robustness of this scheme [22]. In 2021, Muhammad Asghar Khan proposed hyperelliptic curve cryptography (HECC) techniques which is based on user authentication and privacy [23]. In 2021, Chenyu Wang proposed an efficient privacy-preserving user authentication scheme which works on sensor nodes. This scheme has highly secured and less computational costs [24]. In 2020, Sunakshi Singh introduced lightweight mutual authentication schemes for fog nodes. They work on the concept of identity-based encryption [25]. In 2021, Merzougui Salah Eddine proposed blockchain-based secure authentication scheme against node replication attack, wormhole attack, distributed denial-of-service attack, etc. [26]. In 2021, a user authentication scheme has been proposed by Damandeep Kaur. This scheme is more reliable and secure compared to other traditional techniques [27]. In 2021, Ya-Fen Chang proposed three-factor authentication schemes. They use hash function as well as NOR operations. This scheme ensures security, privacy, confidentiality as well as an integrity [28]. In 2020, Chi-Tung Chen proposed an efficient and secure authentication scheme by using wireless sensor network. This scheme has superior performance in terms of efficiency, cost as well as consumption [29]. In 2021, Diksha Rangwani proposed a privacy-preserving secure authentication scheme for wireless sensor network. They used AVISPA simulation tools to verify the accuracy of the system [30]. In 2019, Mohammed EI-hajj reviewed about internet of things (IoT)-based authentication field. It compares various authentication schemes, its strength, weakness and concludes the result [31]. In 2017, Ashwini Bhore proposed a secure object tracking authentication scheme which is based on radio frequency identification system. Simulation result shows that this scheme has more secured features which is better comparison to other authentication schemes [32]. In 2022, Cong Dai proposed a password protection scheme based on elliptic curve cryptography. This scheme reduces password changing and guessing attempts. It provides higher level of security [33]. In 2017, Sunil Kumar reviewed about various authentication schemes which are used for vehicular ad-hoc network. They discuss the strength, weakness, outcome, and future scope of such techniques [34].

Our contributions in this chapter are summarized as follows:

- Reviewed about various authentication schemes which are used for maintaining the confidentiality, integrity, and security of information. Motivation behind this work is to analyze the shortcomings which are in different authentication schemes so that we can implement a better scheme which has more secured and reliable features.
- Highlight benefits of network resource management in VCN.
- Conclusion of the chapter.

13.3 Importance of Vehicular Communication Network

VCN is important because it establishes communication between different vehicles. It extends visibility of road and disseminates safety information. In this network, vehicles are able to exchange their information. Vehicles equipped with appropriate software can utilize the message from another vehicle. The main objective of establishing VCN is to avoid the accidental condition. VCN is usually operated in highway traffic scenario as well as in city. In highway traffic scenario, the environment is simple, whereas in city, the environment is complex. In VCN, vehicles are constrained by road topology and have to respond to other moving vehicles, leading to predictability in terms of their mobility. VCN is very effective for traffic management, assisting drivers as well as improving fuel efficiency. VCN prevents possible crashes. We can minimize up to 70 to 80% accidents through VCN. From Figure 13.1, we can see that there is ENodeB which is used for +- communication between vehicles. Through ENodeB, vehicles can communicate with each other.

There are several advantages in using VCN:

- *Efficient traffic management*: Through VCN, we can track the location of the vehicle, identify traffic area, and find the shortest route for reaching the destination (Figure 13.2).
- *Driver assistance*: VCN instructs drivers about shortest path for reaching the destination to help drivers about unsafe drift.
- *Security*: V2V communication is highly secure because there are many schemes, which validate the authenticity of the message.
- *Privacy*: VCN processes private data about drivers. This data will be highly secured and not tracked by anyone.
- *Safety*: VCN provides safety by assisting drivers at right time. It reduces accidents as well as improves efficiency of drivers.
- *Efficient fuel consumption*: In VCN, all vehicles have to adjust their speed and position so that they can communicate effectively with each other. It saves fuel and other resources.
- *Efficient route optimization*: In VCN, all vehicles can get necessary information about route, traffic, source, destination, etc.
- *Preventing possible crashes*: VCN can prevent possible crashes and reduces 70 to 80 % chances of accident.

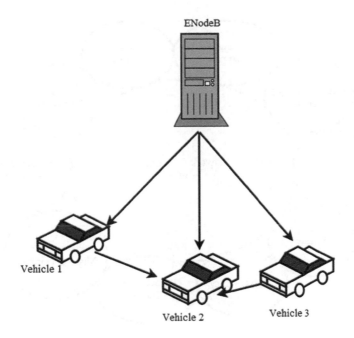

FIGURE 13.1
Vehicular communication network.

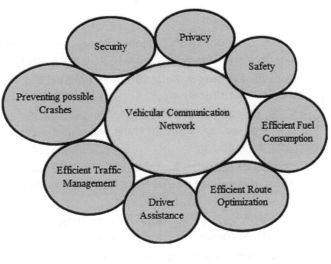

FIGURE 13.2
Advantages of VCN.

13.4 Benefits of Network Resource Management in Vehicular Communication Network

Resources are needed for efficient vehicular communication. They are responsible for whole process of communication. In VCN, resources can be sensor, transmitter, receiver, or many other. Network resource management has the following benefits (Figure 13.3).

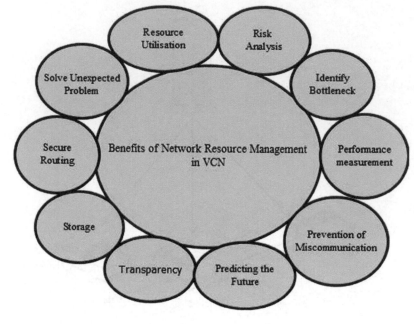

FIGURE 13.3
Benefits of network resource management in VCN.

- *Resource utilization*: Resource utilization means proper use of resources. It means resources are used to their maximum potential to provide response on time.
- *Solve unexpected problem*: Resources help to solve unexpected problems which occur during vehicular communication.
- *Secure routing*: In VCN, messages need to be delivered to specific areas. Resources worked well for message passing. Without transmitter and receiver, it will not be possible to exchange information.
- *Storage*: It is always hard to keep track of the details at one place. Resources help to store all information in database so that the user can retrieve information when they need.
- *Transparency*: Resources provide better view of everything, giving better visibility and clear the flow of whole communication process.
- *Predicting the future*: Resource management helps us to predict the result of using resources in advance.
- *Risk analysis*: Risk analysis involves examining the changes that occur in outcomes due to the impact of risk event.
- *Identify bottleneck*: Resource management help us to identify the bottleneck in advance so that we can fix it.
- *Performance measurement*: Resource management is a continuous activity. After planning, it shows the clear picture of whole process.
- *Preventing miscommunication*: Every resource is assigned to their task. Therefore, there are less chances of any confusion or misunderstanding.

13.5 Requirements of Security in Resource Management

- *Timely verification*: In this authentication mechanism, the receiver has to check that the message is sent by a valid vehicle or not. The receiver also checks that the message has not been changed in the process of transmission. It maintains the confidentiality of the message (Figure 13.4).

- *Non-repudiation*: It provides permission to a third party, which works for generating messages between the sender and receiver.

- *Denial-of-service resistant*: Here, we can verify the invalid signature easily. If an unauthorized person can send invalid signature repeatedly, then there must be chances of denial-of-service attack.

- *Packet loss*: Packet loss is a normal process in wireless sensor network. If there is a packet loss during the process of transmission, the mechanism should verify the next packet.

- *Attack detection*: Attack detection is required to avoid unauthorized access. It is needed to maintain confidentiality of information.

- System failure: Back up is the main part of security. If system fails, we have to recover the data through database or other resources.

13.6 Network Resource Management Authentication Scheme

There are different authentication schemes for network resource management in VCN. These techniques prevent messages from different types of security attack and maintain the confidentiality of information (Figure 13.5).

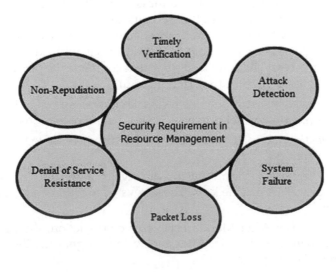

FIGURE 13.4
Requirement of security in network resource management.

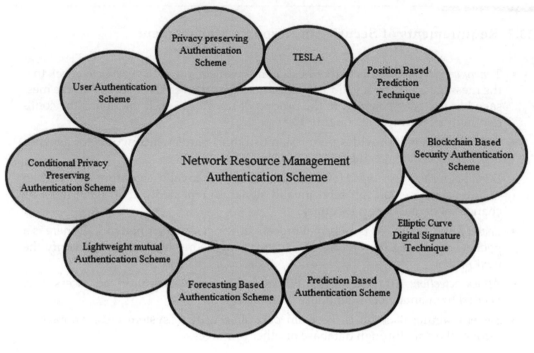

FIGURE 13.5
Authentication schemes of network resource management.

i. *Forecasting-based authentication scheme*: This scheme is based on symmetric cryptography. It improves the efficiency of an authentication. It minimizes the computational charges of an authentication. It analyzes the attack which occurs in vehicular communication process. An efficient authentication scheme should deliver a timely message. It validates the authenticity of the message during transmission process. This technique keeps shortened message so that it cannot be modified by denial-of-service (DoS) attacks or other security threats.

ii. *Prediction-based authentication scheme*: This scheme is basically used to authenticate beacons. It includes sender and receiver. Signature is generated by the sender and verified by the receiver. There are four steps that complete the process of signature generation to signature verification. (Figure 13.6)

   o *Chained key generation*: In this step, each vehicle generates a private key for authentication. This private key helps to analyze the arrival of beacons in advance.

   o *Location prediction*: At each attack, the vehicle predicts its location during the arrival of next attack.

   o *Merkle Hash tree construction*: After predicting location, the vehicle has to use public keys as well as private keys. These keys tie together and generate prediction outcome.

   o *Signature generation*: After Merkle Hash tree construction, the process of generation of signature will begin. After generating a signature, the vehicle has to perform the following steps:

   o *Media access control storage*: To decrease the costs of an unverified signature, the receiver only records a short message. Here, PBA provides security of message.

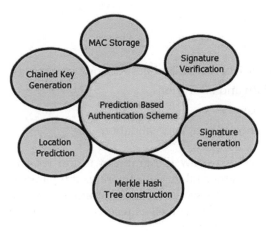

FIGURE 13.6
Prediction-based authentication scheme.

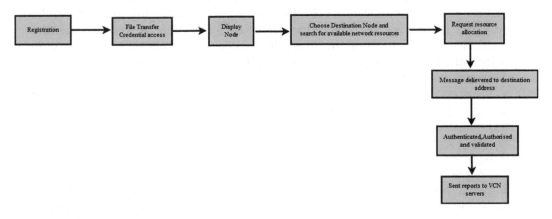

FIGURE 13.7
PBA authentication process.

- *Signature verification:* After the process of MAC storage receiver, verify the signature with the help of TESLA key.
- Figure 13.7 illustrates about PBA authentication scheme.
- The user has to register for the system. If the user is already registered, then the user has to log in the system with their ID and password.
- Display nodes which represent vehicle (nodes are with their unique ID).
- Select source and destination nodes for communication.
- Select file, which has details about vehicles.
- Separate file, which has details about different kinds of attacks.
- Now, we have to split both files. One file has vehicle details and others have details about security attack.
- Set forwarding mode, which provides information about an upcoming attack.
- On the receiver side, calculate bandwidth and validate result data.
- Calculate time, which is incurred in the whole process.

Pseudocode for PBA scheme:

- Registration of vehicle
- Display nodes and select source and destination vehicle for communication
- Calculate bandwidth and allocation
- Send message to receiver
- Authentication, authorization, and validation

After time calculation, we can draw the graph of it and we can measure the accuracy and performance of this scheme. Table 13.1 illustrates the parameters, which are used for prediction, based authentication scheme. These parameters can be used to calculate the computational cost of both sender as well as receiver side. Table 13.2 illustrates the comparative study of different techniques.

- *Elliptic curve digital signature technique*: This security scheme is used to predict and detect the abnormal nodes from the route. This technique identifies those nodes that interrupt communication between vehicles and isolate those nodes from the network. It includes key generation, signature generation as well as signature verification. Key generation includes primary key generator, signature generation is for validating message, and signature verification is used for abnormal node detection. It maintains the ratio about number of packets sent successfully as well as the number of dropped packets. This technique assures node availability and integrity and improves network performance. It offers reliable vehicular communication. This scheme enhances throughput as well as abnormal node detection in the VCN (Figure 13.8).

  Blockchain-based security authentication scheme: IoT has been widely used in various applications such as education, transportation, and military. One of the challenging tasks in IoT is to identify authentication of device. This scheme verifies the authenticity of a device with square root algorithm. This scheme offers more accurate outcomes compared to other authentication schemes. In this scheme, here is a master node that is responsible for whole authentication process. This node verifies the device through a unique no and sends the message to the user about device authentication. From Figure 13.9, we see that there is a blockchain network. There is a master node which is responsible for vehicle authentication and send the message to the relevant user.

  Position-based prediction technique: This technique predicts the position of attacked vehicle. In this approach, we can identify the location of attackers who interrupt VCNs. This scheme blocks attacked nodes from network so that this

TABLE 13.1

Parameters Used for PBA.

Parameters	Description
Sender's computational overhead	Average time for signature generation
Receiver's computational overhead	Average time for signature verification
Packet processing rate	Verification security of attack
Storage cost for verification of beacon	Amount of bytes stored by vehicle

TABLE 13.2

Comparative Study of Different Network-based Authentication Techniques.

Authors	Techniques	Limitations	Accuracy Rate (%)
Vinod Namboodiri et Al. [1]	Prediction-based routing protocol	This protocol does not work well for power control.	95
Chen Lyu et al. [3]	Prediction-based authentication scheme	This technique does not fulfill privacy and security requirement.	99
Harmanjot Kaur et al. [6]	Clustering-based routing approaches	Less throughput and high packet drop ratio.	85
Sami Abduljabbar Rashid et al. [8]	Cluster in a wireless sensor network	It is not suitable for small-scale wireless network.	90
Zongjian He et al. [16]	SDN-based vehicular communication Network	Lack of scalability and security.	80
P. Deepalakshmi et al. [17]	Elliptic curve digital signature technique	Less throughput and higher packet drop rate ratio.	90
Chenyu Wang et al. [24]	Privacy preserving authentication scheme	It works only for sensor nodes	80
Sunakshi Singh et al. [25]	Lightweight Mutual authentication scheme	Less secured and high computational costs	88
Merzoungi et al. [26]	Blockchain-based secure authentication Scheme	High message drop ratio	75
Damandeep Kaur et al. [27]	User authentication scheme	Less reliable	80

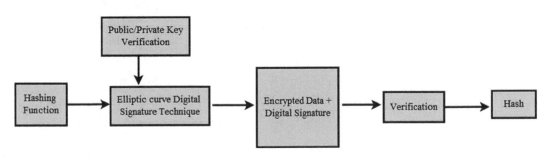

FIGURE 13.8
Elliptic curve digital signature technique.

node never interrupts communication again and this process works smoothly. This technique may fail when there are multiple attacks at the same time. This condition is called blockage and this is the limitation of this technique. In this technique, we can predict the position of attackers in advance so that we can avoid interruption in VCN (Figure 13.10).

- *TESLA*: TESLA stands for timed efficient stream loss-tolerant authentication protocol. It is used for message exchange between vehicles in vehicular communication network. It sends only a short message authentication code and maintains the security of information. The receiver decodes the message, reads it, and

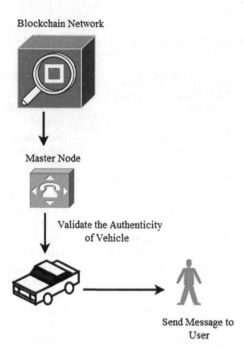

FIGURE 13.9
Blockchain-based security authentication scheme.

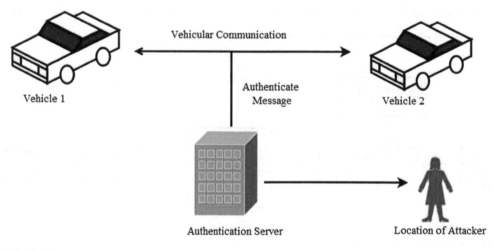

FIGURE 13.10
Position-based prediction technique.

replies back with a short message so that any unauthorized user cannot access it (Figure 13.11).

Privacy-preserving authentication scheme: It is based on public as well as private key. Public key is used by the sender for message encryption and private key is used by the receiver for message decryption. Public keys and private keys are responsible for authentication and integrity. This scheme is basically used to prevent hackers from modifying the message. For this scheme, it is compulsory for all vehicles to be equipped with on-board unit (OBU) and sensor unit (Figure 13.12).

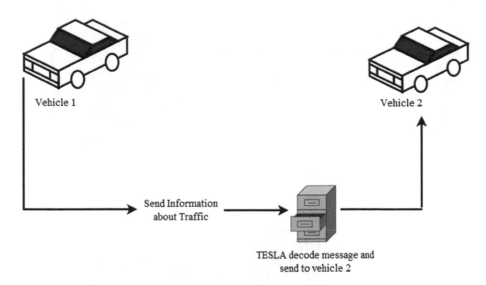

FIGURE 13.11
TESLA authentication technique.

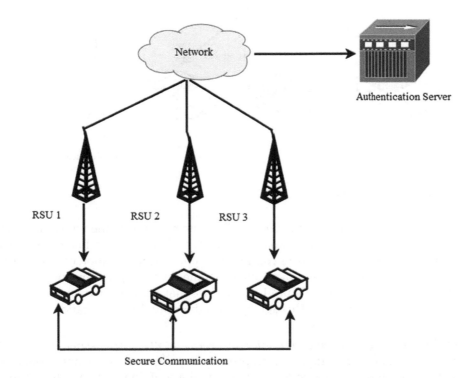

FIGURE 13.12
Privacy-preserving authentication scheme.

- *User authentication scheme*: In this scheme, the authenticity of user has been done by smart devices.
  These devices check that the message has been successfully received by the receiver or not. This scheme is basically used for the verification of user (Figure 13.13).

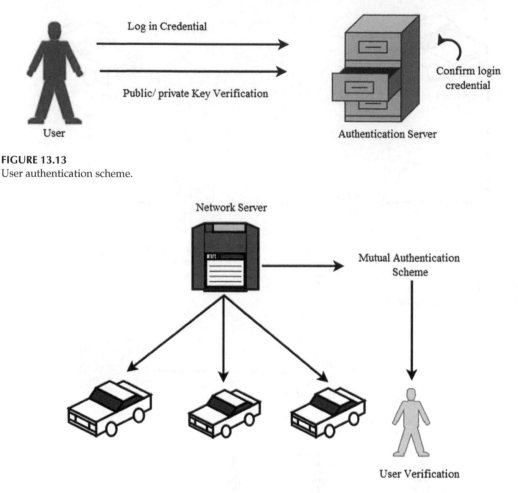

FIGURE 13.13
User authentication scheme.

FIGURE 13.14
Lightweight mutual authentication scheme.

- *Lightweight mutual authentication scheme*: This scheme protects the whole network from an unauthorized access. It enables mutual authentication schemes between two parties. This scheme reduces computation and communication complexities. This scheme is robust against attacks and consumes less computation and communication energy costs. From Figure 13.14, we see that there is a network in which three vehicles are connected. Through lightweight mutual authentication schemes, user verification has been done.

- *Conditional privacy-preserving authentication scheme (CPPA)*: This scheme is based on the concept of cryptography. Here, primary key, signature assistant node, task allocation, and tokens are used for maintaining the confidentiality of information. This scheme is safe and secure and able to save integrity of information from an unauthorized source.

- Steps of CPPA schemes:
  i. Primary and secondary signature – It takes the length of the message as input and converts it into the shortest form for maintaining the secrecy of an information (Figure 13.15).

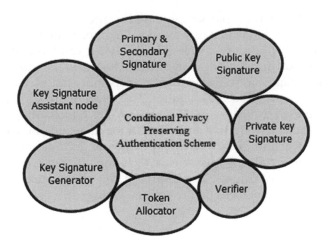

FIGURE 13.15
Conditional privacy-preserving authentication scheme.

ii. Key signature assistant node – It is a generated by vehicles before sending an information to the nearby vehicle. This is basically used to verify and validate the authenticity of the message. The receiving vehicle just validates the signature and takes a decision about the authenticity of a message.

iii. Key signature generator – After receiving an information, the receiving vehicle generates signature and sends it with a confirmation message to the sender. The sender validates the signature and assures that the message has been delivered to the destination.

iv. Token allocator – When the vehicle has to be registered, a token will be allocated. This token will verify the license of the vehicle. If any vehicle does not have a token, it has to be out from vehicular communication process.

v. Verifier – Verifier checks the validity of the message. The verifier checks the whole process from sender to receiver and validate the confidentiality of message.

vi. Public key, private key signature (PKS) – The sender attaches a public key with the message. The sender sends the message in an encrypted form. The receiver decrypts the message using a private key.

13.7 Conclusion and Future Scope

In this chapter, we have examined a variety of authentication systems that can help us to predict and identify attackers ahead of time. Signature Generation, MAC Storage, and signature verification are the three primary steps of the PBA Scheme. This scheme's key benefit is that it is safe and secure. The main disadvantage of this method is that it does not address the issue of privacy. PBA keeps the messages in it private and confidential. It is not accessible to anyone who is not authorized. Elliptic curve digital signature technique protects unauthorized users from modifying the confidential information. It is more secure but has a higher packet drop ratio. Blockchain-based secure authentication scheme protects information from different kinds of cyber-attacks. Such techniques have a high message drop ratio. The user authentication scheme is basically for validating the user. It uses the public key and private key for identification of users. For future, we need to improve such techniques. Every technique has some security and reliability issues. We need to overcome such issues and try to improve such techniques. These technique are very effective, light-weighted, and can be used to analyze packet loss. In VCN, a forecasting-based authentication technique is particularly effective and efficient for managing network resources. We can improve this approach in the future by providing an accurate prediction model. We can use machine learning and cognitive science technique for better metric value. We can also think about the issue of privacy. This approach can be implemented in a real-world context. In the future, we must focus on the privacy and security standards, as well as the accuracy ratio.

References

1. Namboodiri, V., & Gao, L. (2007). Prediction-based routing for vehicular ad hoc networks. *IEEE Transactions on Vehicular Technology*, 56(4), 2332–2345.
2. Khairnar, V. D., & Pradhan, S. N. (2010). Comparative study of simulation for vehicular ad- hoc network. *International Journal of Computer Applications*, 4(10), 15–18.
3. Lyu, C., Gu, D., Zeng, Y., & Mohapatra, P. (2015). PBA: Prediction based authentication for vehicle-to-vehicle communications. *IEEE Transactions on dependable and Secure Computing*, 13(1), 71–83.
4. Al Mamun, A., Sohel, M., Mohammad, N., Sunny, M. S. H., Dipta, D. R., & Hossain, E. (2020). A comprehensive review of the load forecasting techniques using single and hybrid predictive models. *IEEE Access*, 8, 134911–134939.
5. Evermann, J., Rehse, J. R., & Fettke, P. (2017). Predicting process behaviour using deep learning. *Decision Support Systems*, 100, 129140.
6. Vehicular, R. A. I. (2018). A review of clustering based routing appproaches in vehicular adhoc network. *International Journal of Advanced Research in Computer Engineering & Technology (IJARCET)*, 7(2). DOI: 10.17485/ijst/2015/v8i1/106884
7. Masdari, M., & Khoshnevis, A. (2020). A survey and classification of the workload forecasting methods in cloud computing. *Cluster Computing*, 23(4), 2399–2424.
8. Hamdi, M. M., Audah, L., Rashid, S. A., Mohammed, A. H., Alani, S., & Mustafa, A. S. (2020, June). A review of applications, characteristics and challenges in vehicular ad hoc networks (VANETs). In *2020 International Congress on Human-Computer Interaction, Optimization and Robotic Applications (HORA)* (p. 17). IEEE.
9. Jaiswal, R. K. (2020). Position-based routing protocol using Kalman filter as a prediction module for vehicular ad hoc networks. *Computers & Electrical Engineering*, 83, 106599.
10. Agarwal, Vartika, Sharma, Sachin, & Agarwal, Piyush. "IoT Based Smart Transport Management and Vehicle-to-Vehicle Communication System." In *Computer Networks, Big Data and IoT*, pp. 709–716. Springer, Singapore, 2021.
11. Agarwal, Vartika, & Sharma, Sachin. "IoT based smart transport management system." In *International Conference on Advanced Informatics for Computing Research*, pp. 207–216. Springer, Singapore, 2020.
12. Mansoor, M., Grimaccia, F., Leva, S., & Mussetta, M. (2021). Comparison of echo state network and feed-forward neural networks in electrical load forecasting for demand response programs. *Mathematics and Computers in Simulation*, 184, 282–293.

13. Agarwal, Vartika, Sharma, Sachin, & Bansal, Gagan. "Secured Scheduling Techniques of Network Resource Management in Vehicular Communication Networks." In *2021 5th International Conference on Intelligent Computing and Control Systems (ICICCS)*, pp. 198–202. IEEE, 2021.

14. Agarwal, Vartika, & Sharma, Sachin (2022). "Deep Learning Techniques to Improve Radio Resource Management in Vehicular Communication Network", In *International Conference on Sustainable Advanced Computing (ICSAC 2021)*.

15. Liu, H., Li, Y., Duan, Z., & Chen, C. (2020). A review on multiobjective optimization framework in wind energy forecasting techniques and applications. *Energy Conversion and Management*, 224, 113324.

16. He, Z., Cao, J., & Liu, X. (2016). SDVN: Enabling rapid network innovation for heterogeneous vehicular communication. *IEEE Network*, 30(4), 10–15.

17. Deepalakshmi, P., & Kumanan, T. (2020, September). Elliptic curve digital signature technique based abnormal node detection in wireless ad hoc networks. In *IOP Conference Series: Materials Science and Engineering* (Vol. 925, No. 1), p. 012075. IOP Publishing.

18. Liu, C. H., & Chung, Y. F. (2017). Secure user authentication scheme for wireless healthcare sensor networks. *Computers & Electrical Engineering*, 59, 250–261.

19. Irshad, A., Chaudhry, S. A., Alazab, M., Kanwal, A., Zia, M. S., & Zikria, Y. B. (2021). A secure demand response management authentication scheme for smart grid. *Sustainable Energy Technologies and Assessments*, 48, 101571.

20. Hussain, S., Chaudhry, S. A., Alomari, O. A., Alsharif, M. H., Khan, M. K., & Kumar, N. (2021). Amassing the security: An ECC-based authentication scheme for Internet of drones. *IEEE Systems Journal*, 15, 4431–4438.

21 Sadhukhan, D., Ray, S., Biswas, G. P., Khan, M. K., & Dasgupta, M. (2021). A lightweight remote user authentication scheme for IoT communication using elliptic curve cryptography. *The Journal of Supercomputing*, 77(2), 1114–1151.

22. Shariq, M., Singh, K., Maurya, P. K., Ahmadian, A., & Ariffin, M. R. K. (2021). URASP: An ultra-lightweight rfid authentication scheme using permutation operation. *Peer-to- Peer Networking and Applications*, 14, 1–21.

23. Khan, M. A., Ullah, I., Alkhalifah, A., Rehman, S. U., Shah, J. A., Uddin, I. I., … and Algarni, F. (2021). A provable and privacy-preserving authentication scheme for UAV- enabled intelligent transportation systems. *IEEE Transactions on Industrial Informatics*, 18, 3416–3425.

24. Wang, C., Wang, D., Xu, G., & He, D. (2022). Efficient privacy-preserving use authentication scheme with forward secrecy for industry 4.0. Science China. *Information Sciences*, 65(1), 1–15.

25. Singh, S., & Chaurasiya, V. K. (2021). Mutual authentication scheme of IoT devices in fog computing environment. *Cluster Computing*, 24(3), 1643–1657.

26 Eddine, M. S., Ferrag, M. A., Friha, O., & Maglaras, L. (2021). EASBF: An efficient authentication scheme over blockchain for fog computing-enabled internet of vehicles. *Journal of Information Security and Applications*, 59, 102802.

27. Kaur, D., & Kumar, D. (2021). Cryptanalysis and improvement of a two-factor user authentication scheme for smart home. *Journal of Information Security and Applications*, 58, 102787.

28. Chang, Y. F., Tai, W. L., Hou, P. L., & Lai, K. Y. (2021). A secure three-factor anonymous user authentication scheme for internet of things environments. *Symmetry*, 13(7), 1121.

29. Chen, C. T., Lee, C. C., & Lin, I. C. (2020). Efficient and secure three-party mutual authentication key agreement protocol for WSNs in IoT environments. *PloS One*, 15(4), e0232277.

30. Rangwani, D., Sadhukhan, D., Ray, S., Khan, M. K., & Dasgupta, M. (2021). An improved privacy preserving remote user authentication scheme for agricultural wireless sensor network. *Transactions on Emerging Telecommunications Technologies*, 32(3), e4218

31. El-Hajj, M., Fadlallah, A., Chamoun, M., & Serhrouchni, A. (2019). A survey of internet of things (IoT) authentication schemes. *Sensors*, 19(5), 1141.

32. Bhore, M. A., & Bhandari, G. *An analysis of RFID authentication schemes with Secure Object Tracking Protocol for the Internet of Things*.

33. Dai, C., & Xu, Z. (2022). A secure three-factor authentication scheme for multi-gateway wireless sensor networks based on elliptic curve cryptography. *Ad Hoc Networks*, 127, 102768.
34. Manvi, S. S., & Tangade, S. (2017). A survey on authentication schemes in VANETs for secured communication. *Vehicular Communications*, 9, 19–30.

14

Intelligent Cyber-Physical Systems Security for Industry 4.0: Concluding Remarks

Jyoti Sekhar Banerjee

Bengal Institute of Technology, Kolkata, India

Siddhartha Bhattacharyya

Rajnagar Mahavidyalaya, Birbhum, India

CONTENTS

14.1 Introduction

With the advent of electronic sensors and actuators driving almost every aspect of today's technology, there has been a rising concern about data breaches, data compromise as well as theft of data across platforms. Intruders have become more diligent than ever before in intercepting different data-intensive platforms and applications leading to the vulnerability of the associated cyber-physical systems [10] which play a profound role in different critical infrastructural and logistic systems of everyday lives. As such ensuring and envisaging proper security of the underlying systems has assumed paramount importance.

Cyber-physical systems security, as the name suggests, refers to the security concerns related to these physical systems used to envisage cybersecurity solutions, [15–16] which include internet of things (IoT), [11–14] industrial internet of things (IIoT), operational technology, and industrial control systems to name a few [1–4].

Cyber-physical systems entail smart networked systems comprising embedded sensors, high-speed multi-data processors, and physical actuators for sensing and interacting with the physical world or environment, thereby ensuring seamless real-time operations of the intended infrastructure and logistics. These technologies are relatively cheaper and can be managed remotely, thanks to the advancement in sensing and networking technologies. Moreover, the deployment of these systems not only lowers the associated costs but also induces a hassle-free and minimal maintenance system.

However, since these systems are distributed, networked, and require operating systems and applications to process the associated data, these systems are prone to attacks by intruders and interceptors. The obvious consequences of such malicious cyber and

DOI: 10.1201/9781003241348-14

physical attacks may lead to catastrophic disasters. Hence, cyber-physical security, which is intended to protect and safeguard these resources outside the ambit of traditional information technology, takes into cognizance several attributes, including communication protocols, environmental conditions, security audits coupled with security management strategies to ensure continuous, and reliable operations.

14.2 Requisite Attributes

Several unique attributes require proper modeling and consideration for ensuring cyber-physical security [5–8]. These can be listed as follows:

Physical access: Although it looks to be a simple requirement, one should ensure robust physical access to the network ports, network cables, network connections, and even power connections to prevent tampering with the attached physical devices. This can be incorporated by resorting to physical key locks, tamper-proof screws, encapsulated cables, and deploying hardwired devices.

Environmental conditions: One of the most important concerns with the smooth operation of the cyber-physical systems lies in the protection of these devices from environmental hazards in the form of extreme temperature, moisture, sand/dust, and other corrosive elements. Hence, the devices need to be tested under extreme environmental conditions before they are put to use in real-time operations.

Cybersecurity: As these devices are essentially distributed networked systems, they are prone to any electronic flaws that figure in any other similar systems or applications. Prospective threats in this regard can be avoided by passing the inherent firmware components through different security checks, including penetration testing, vulnerability assessments, change control, and patch management. This also includes proper physical manipulation of external devices controlled by these resources as they can pose equal threats to human lives and the physical environment.

Planning: Since these devices are specially meant for managing the security of cyber-physical systems, the corresponding operating procedures and planning are ought to be unique in all respects. Hence, it is not advisable to carry out active penetration tests during normal operations. Simple troubleshooting processes are also not suitable for these devices as any attempt for remediation of an identified fault in these operational systems may lead to improper functioning of other system modules. Thus, one needs to have proper operational backup plans, processes for outages, and exception handling mechanisms for ensuring uninterrupted services.

Risk analysis: Because these devices require interaction with the physical world, a proper assessment of associated risks is required at each operational layer to estimate the potential impact of the risks and corresponding mitigation strategies in order to prevent any catastrophic disaster.

Privacy: These data-intensive systems deal with lots of personally identifiable information retrieved from recorders to cameras to card readers. Due to the sensitivity of these data, proper encryption, secured storage, and privacy of these data must be ensured to protect private information.

Reliability: The primary attribute of any cyber-physical system is its reliability of operation. Thus, frequent rebooting of these systems is not at all recommended since this would hamper the seamless operation of these devices. Hence, management and redressal of power faults and recovery of other disasters should be scheduled outside the runtime period so that the normal operations of these systems are not compromised.

Supportability: Given the wear and tear of the system components with time, one of the most important attributes, which requires proper attention, is the supportability of all the components in these systems. Supportability refers to the support for continuous lifetime availability of the spare parts for the physical components as well as the support for software upgrades and security patches. Moreover, these systems being highly scalable, means for uninterrupted up-gradation of the system components should be also ascertained. Otherwise, the system may become non-operational before the planned depreciation period.

14.3 Role of Computational Intelligence in Cyber-Physical Security Systems

Almost all the discussed attributes require manual intervention and human intelligence for ensuring proper management and assessment. As such, the performance of each attribute is subject to human error and understanding. Fully automated and intelligent management of these attributes is therefore needed to prevent manual intervention. Computational intelligence methods [9, 17–18] may come in good stead in this regard, thanks to the evolution of intelligent tools and techniques. All the attributes may thus be further guided by intelligent methods to obviate manual interpretation and understanding, thereby providing a robust and failsafe alternative to the traditional attribute assessment procedures, while taking into cognizance the uncertainty and imprecision in human observations. [19–21]

The bottom line is therefore the inculcation of computational intelligence into the security mechanisms, making it knowledgeable and self-reliant.

References

1. Kim, N. Y., Rathore, S., Ryu, J. H., Park, J. H., & Park, J. H. (2018). A survey on cyber physical system security for IoT: issues, challenges, threats, solutions. *Journal of Information Processing Systems*, 14(6), 1361–1384.
2. Lee, J., Bagheri, B., & Kao, H. A. (2015). A cyber-physical systems architecture for industry 4.0-based manufacturing systems. *Manufacturing Letters*, 3, 18–23.
3. Huang, S., Zhou, C. J., Yang, S. H., & Qin, Y. Q. (2015). Cyber-physical system security for networked industrial processes. *International Journal of Automation and Computing*, 12(6), 567–578.
4. Wurm, J., Jin, Y., Liu, Y., Hu, S., Heffner, K., Rahman, F., & Tehranipoor, M. (2016). Introduction to cyber-physical system security: A cross-layer perspective. *IEEE Transactions on Multi-Scale Computing Systems*, 3(3), 215–227.

5. Gupta, R., Tanwar, S., Al-Turjman, F., Italiya, P., Nauman, A., & Kim, S. W. (2020). Smart contract privacy protection using ai in cyber-physical systems: Tools, techniques and challenges. *IEEE Access*, 8, 24746–24772.

6. Stefanov, A., & Liu, C. C. (2014). Cyber-physical system security and impact analysis. *IFAC Proceedings Volumes*, 47(3), 11238–11243.

7. Oks, S. J., Jalowski, M., Fritzsche, A., & Möslein, K. M. (2019). Cyber-physical modeling and simulation: A reference architecture for designing demonstrators for industrial cyber-physical systems. *Procedia CIRP*, *84*, 257–264.

8. Alam, K. M., & El Saddik, A. (2017). C2PS: A digital twin architecture reference model for the cloud-based cyber-physical systems. *IEEE Access*, *5*, 2050–2062.

9. Bhattacharyya, S., Maulik, U. & Bandyopadhyay, S. (eds. Dai, Y., Chakraborty, B., & Shi, M.). Soft Computing and its Applications. *Kansei Engineering and Soft Computing: Theory and Practice*, IGI Global, Hershey PA, USA, pp. 1–30, 2011.

10. Das, K., & Banerjee, J. S. (2022). Green IoT for Intelligent Cyber-Physical Systems in Industry 4.0: A Review of Enabling Technologies, and Solutions. In *GCAIA 21: Proceedings of 2nd Global Conference on Artificial Intelligence and Applications* (pp. 463–478).

11. Das, K., & Banerjee, J. S. (2022). Cognitive Radio-Enabled Internet of Things (CR-IoT): An Integrated Approach towards Smarter World. In *GCAIA 21: Proceedings of 2nd Global Conference on Artificial Intelligence and Applications*, 541–555.

12. Biswas, S., Sharma, L. K., Ranjan, R., Saha, S., Chakraborty, A., & Banerjee, J. S. (2021). Smart farming and water saving-based intelligent irrigation system implementation using the internet of things. In *Recent Trends in Computational Intelligence Enabled Research* (pp. 339–354). Academic Press.

13. Roy, R., Dutta, S., Biswas, S., & Banerjee, J. S. (2020). Android Things: A Comprehensive Solution from Things to Smart Display and Speaker. In *Proceedings of International Conference on IoT Inclusive Life (ICIIL 2019)*, NITTTR Chandigarh, India (pp. 339–352). Springer, Singapore.

14. Pandey, I., Dutta, H. S., & Banerjee, J. S. (2019, March). WBAN: A Smart Approach to Next Generation e-Healthcare System. In *2019 3rd International Conference on Computing Methodologies and Communication (ICCMC)* (pp. 344–349). IEEE.

15. Banerjee, J., Maiti, S., Chakraborty, S., Dutta, S., Chakraborty, A., & Banerjee, J. S. (2019, March). Impact of Machine Learning in Various Network Security Applications. In *2019 3rd International Conference on Computing Methodologies and Communication (ICCMC)* (pp. 276–281). IEEE.

16. Chakraborty, A., Banerjee, J. S., & Chattopadhyay, A. (2020). Malicious node restricted quantized data fusion scheme for trustworthy spectrum sensing in cognitive radio networks. *Journal of Mechanics of Continua and Mathematical Sciences*, 15(1), 39–56.

17. Chakraborty, A., Singh, B., Sau, A., Sanyal, D., Sarkar, B., Basu, S., & Banerjee, J. S. (2022). Intelligent Vehicle Accident Detection and Smart Rescue System. In *Applications of Machine Intelligence in Engineering* (pp. 565–576). CRC Press.

18. Chattopadhyay, J., Kundu, S., Chakraborty, A., & Banerjee, J. S. (2018, November). Facial Expression Recognition for Human Computer Interaction. In *International Conference on Computational Vision and Bio Inspired Computing* (pp. 1181–1192). Springer, Cham.

19. Banerjee, J. S., Chakraborty, A., & Chattopadhyay, A. (2021). A decision model for selecting best reliable relay queue for cooperative relaying in cooperative cognitive radio networks: the extent analysis based fuzzy AHP solution. *Wireless Networks*, 27(4), 2909–2930.

20. Chakraborty, A., Banerjee, J. S., & Chattopadhyay, A. (2019). Non-uniform quantized data fusion rule for data rate saving and reducing control channel overhead for cooperative spectrum sensing in cognitive radio networks. *Wireless Personal Communications*, 104(2), 837–851.

21. Banerjee, J. S., Chakraborty, A., & Chattopadhyay, A. (2022). A cooperative strategy for trustworthy relay selection in CR network: a game-theoretic solution. *Wireless Personal Communications*, 122(1), 41–67.

Index